Teaching Science so that Students *Learn Science*

A Paradigm for Christian Schools

By
John D. Mays

Copyright © 2009, John D. Mays

All rights reserved. No part of this book may be reproduced or transmitted in any form or by any means, electronic or mechanical, including photocopying, recording, or by information storage and retrieval systems, without the written permission of the publisher, except by a reviewer who may quote brief passages in a review.

Scripture quotations are from The Holy Bible, English Standard Version, copyright ©2001 by Crossway Bibles, a publishing ministry of Good News Publishers. Used by permission. All rights reserved.

Published by
Novare Science and Math
P.O. Box 92934
Austin, Texas 78709-2934
novarescienceandmath.com

Printed in the United States of America

ISBN-13: 978-0-615-33599-5

Library of Congress Control Number: 2009940517

Dedication

This book is dedicated to my mother. She is an intelligent lady who has never heard about the things I am discussing here, so I reckoned that if the book is intelligible to her it would also make sense to those to whom it is written. As she was reading through the manuscript she found all the annoying little grammatical errors I had missed. It now seems clear that through her use of this singular talent she was able to protect me from the absurd hubris of believing I had outstripped my parents. Thus, one of the elements of true conservatism: The knowledge and traditions of the communities that raise us are never left behind. Instead, they inform how we live.

Acknowledgments

I wish to express my sincere thanks to those who helped me sharpen my ideas and improve my manuscript. Chris Corley is a valued colleague. We have discussed these ideas and labored to implement them together for years, and he is a keen reader who provided many suggestions to improve the manuscript. Chuck Evans provided me with very welcome early encouragement when I began putting together the first few Chapters. Dr. Keith Miller put me in touch with some of the other individuals in this list whose feedback was invaluable. Mervin Bitikofer provided very valuable commentary that helped me clarify ideas, improve my terminology, and avoid language that might inadvertently foster some of the very misunderstandings I desire to help remedy. Joan Mays is a fine grammarian whose review of the manuscript was very helpful. Dr. Christina Swan gave me some early encouragement as I was putting together the ideas in the final Chapter. Dr. Chris Mack gave me a great deal of crucial input and spent hours with me in conversation helping me develop a more nuanced understanding of the nature of scientific knowledge. Dr. Loren Haarsma helped me tighten up my ideas, terminology and presentation, and gave me much food for thought for my Chapter on Truth and Facts. Doreen Howell, Mindy Clow, Laurie Bunch and Bonnie Shipley all reviewed the manuscript from the point of view of Grammar School teachers. Topher Ayrhart graciously prepared the graphic design for the Cycle of Scientific Enterprise.

Whatever strengths my book possesses were incalculably improved by comments from these gracious contributors. Naturally, whatever weaknesses the book still possesses are my own responsibility.

Table of Contents

Foreword ix
Introduction xiii

Part I General principles for science education, including what science is and how scientific knowledge works.
    1 Core Principles for Instruction in Science   3
    2 Truth and Facts   16
    3 What Is Science?   27
    4 The Cycle of Scientific Enterprise   32

Part II Striving for integration and mastery, with specific teaching tools for the upper grades.
    5 Verbal Matters   49
    6 Quantitative Matters   57
    7 Science as a Cumulative Discipline   77

Part III Additional critical topics pertaining to your school's science program.
    8 Grammar School and the Science Curriculum   105
    9 Laboratory Work and Lab Reports   114
    10 Making History Work in Your High School Science Class   123
    11 Dealing with Evolution   128

Foreword

IN 1947, TO A SUMMER ASSEMBLY OF LAYPEOPLE ENJOYING LECTURES FROM Britain's cultural brain trust, Dorothy L. Sayers made this incisive comment:

> *Is not the great defect of our education today...that although we often succeed in teaching our pupils "subjects," we fail lamentably on the whole in teaching them how to think: they learn everything, except the art of learning.*

In the last 30 years or so, a phalanx of Christian educators have latched on to Sayers' essay *The Lost Tools of Learning* as a means to reinvigorate American education with the notion that teaching subjects is only as useful to students as the extent to which they remember both the content of their courses *and* the manner in which they learned what they remember.

Unfortunately, were Miss Sayers to return from the hereafter and be asked to prepare the same speech, she would likely be forced to amend the statement above with something like "The defect of our education today is that we have forgotten *both* how to teach subjects *and* how to teach students how to think!"

Nowhere is this more evident today than in the areas of science and mathematics. Having spent twenty years in schools, it strikes me that when it comes to the two uncontested engines of the modern economy—science and math—confusion reigns regarding how best to teach them to modern students. In the past several decades, following the glory days of the space race and the all-out effort to place American school kids ahead of the rest of the world in scientific knowledge, we rank in the bottom half among nations pursuing similar goals.

In Christian schools (and I mostly mean Protestant, evangelical schools), the situation often seems even worse. Textbooks and curriculum guides often use science instruction merely as a means to defend against secular notions of how the world was formed. The irony, of course, is that in our effort to reassert the truth of God's existence and his creative power, we often neglect to give students the tools they need to examine the natural order and the amazing complexity and nuance that God designed into it.

John Mays' book might remedy this situation. He writes from the standpoint of a professional Christian educator who has had, over the years, to overcome deficiencies in his own education, not the least of which is the separation of science from philosophy and serious questions of faith. In part, then, this book is a primer on the Christian basis for "doing science." Mays directly answers the question of why Christian people should take science seriously. He argues, in fact, that properly educated Christians are best equipped to be in the forefront of the conversation about scientific things. His argument is not based in economics, national security, or international test comparisons. Mays is motivated by Christian discipleship, which cannot be divorced from the pursuit of truth, wherever it is found.

But that is just a start. Unlike other books on the subject of science and faith, Mays lays out a plan for teaching science in a Christian school which has the effect of providing students with the knowledge to function expertly in a rapidly evolving, highly complex field of study. And this is no untested theory. I know him well, and he has worked tirelessly over fifteen or more years to bring the essential content of modern science together with an instructional approach that ensures that the most average student can emerge from school with competitive scientific knowledge and skills.

This book is written for teachers of science. If you teach in a Christian school, Mays will help you to bring together the underlying theological issues that arise in science with a method that improves your students' grasp of the discipline and their competence in applying it. Even if you don't teach in a Christian school, this book can provide a roadmap out of the malaise of teaching science with little discernible effect on the majority of your students.

But this is also a book for people who care deeply about Christian education, generally. Mays blends pre-modern ideas about the natural world with disciplined attention to standard scientific practices. He defines what the goals of teaching science *should be*. And he clearly knows what he is talking about.

About eight years ago, I arrived on the campus where John Mays currently teaches. Almost immediately, we began talking about the problems that math and science teachers face and the effect on students. We agreed that it is unacceptable to treat math and science as special categories in which some gifted students achieve mastery of upper level concepts and operations, while others are left out.

This book is partly the result of Mays' leadership of and collaboration with highly skilled colleagues in his department to address this problem and others. Arduous though it has been, they have persevered, and much of what they have accomplished together is documented here. It is worth imitating.

Though it may seem unlikely for such a small work, this book has the potential to introduce a generation of teachers and students to the wonder of the Creation and the joy of learning how it all fits together.

Charles T. Evans
Austin, Texas

Introduction

TO MY KNOWLEDGE, THERE IS NO TREATISE ON HOW EFFECTIVE SCIENCE PEDAGOGY should work in Christian schools. Science teachers in Christian schools often bring a tremendous amount of interest, passion and enthusiasm to their work. Some of the most knowledgeable and passionate teachers I have ever met have been the devoted science and math faculty at Christian schools. But in terms of teaching methodology, many simply teach science the way it was taught to them, the way it is commonly taught throughout the country. One of the consequences of this is that many students complete their high school programs knowing very little about science. Of course, this problem is not at all unique to Christian schools, but represents an educational crisis affecting our entire nation and threatening our technological leadership. This needs to change, and I have written this book in an effort to summarize pedagogical principles that are effective in enabling change to happen in the way our students are taught science and what they are expected to learn.

The school where I teach describes itself as a "classical and Christian" school, which means, among other things, our school recognizes the value of a Trivium-based approach to education, and we teach the classical disciplines of Latin, Logic and Rhetoric. It also means that we recognize that the separation of knowledge into the academic disciplines is a pedagogical convenience. Just because science and literature are studied in separate classes does not mean we should treat knowledge like items at a cafeteria, where we take what appeals to us and skip the rest. An educated person should be well versed in both science and literature (and art, mathematics, history, and the rest).

Realizing this in the context of a school science course would mean that, in addition to knowledge of scientific principles, the student writes well, understands the significance of historical characters and events, is able to apply mathematics, and so on. I believe that these skills should be important to Christian educators regardless of whether or not the school is overtly "classical." The principles of good science teaching and effective science learning are the same. Thus, I hope that this text will be of value to all those in Christian schools who are concerned with science instruction.

Actually, most of the principles discussed here will make science pedagogy effective from any point of view, Christian or otherwise. But since in my development of the nature of scientific knowledge I have appealed directly to Christian theology, I have written this book for Christian educators. If others can profit from my pedagogical advice as well, then I would be humbly pleased to have made a more general contribution to science pedagogy.

The starting point for all of my thinking about education is the fundamental idea that human beings are made in the image of God. This means that our material needs, the things we need to take care of our bodies, are not the most important things for us. The most important things are the things that we need to take care of our souls. As Jesus said, "Seek first the kingdom of God and his righteousness, and all these [material] things will be added to you."

Of course, this idea is at the heart of Christianity, so Christian education must have this truth at its heart as well. Everyone agrees that education needs to serve the needs of man, but modern views of the needs of man tend to center on the economic, and thus on the material. Education that centers on the material will hold that the purpose of secondary education is to help people get good jobs or get into good colleges so they can have comfortable lifestyles and so on. In a Christian view of education, economic well being, while not ignored, is not the focus. Economic well being can take care of itself if a person is well educated. We don't have to make getting a good job the target. Instead, helping people develop into good human beings is the target. This principle is the basis for what has traditionally been called a liberal education, and should be at the heart of Christian education.

This is the reason there is such a strong overlap between Christian schools that promote "classical" education and other Christian schools. Classically, the purpose of education is the development of wisdom and virtue through the pursuit of truth, goodness and beauty. In passages such as Proverbs 4:5-7 and Philippians 4:8 the Bible affirms that Christian education should have these same goals in mind. And, as described above, both classical and Christian conceptions of education understand that knowledge, and our approach to it, should be regarded as an integrated whole.

In this book I will focus specifically on matters pertaining to pedagogy in science. Except in one or two cases, I will not address pedagogical elements that should apply to all teaching such as effective classroom management, writing good tests, using a variety of learning activities to appeal to different learning styles, developing healthy relationships with one's students, explaining things clearly, using an organized approach to writing things on the board, maintaining a continuous awareness of one's students' comprehension of the lesson, directing learning objectives toward all six levels of Bloom's Taxonomy, and so on. All effective teachers do these things.

The double meaning in the title about teaching and learning science is intentional. We want our students to learn science, and not something else which may be calling itself science but which doesn't reflect the actual nature of scientific knowledge. This is a real problem today, and many people do not correctly understand the nature of the scientific enterprise and the facts, theories and methods at the core of it. We also want our students to learn well and take the knowledge from our classes with them to help them be better students, citizens, and human beings. Our task will be to examine the specifics that make science instruction effective so that our students *learn science*.

PART I

*General principles for science education,
including what science is and
how scientific knowledge works.*

Chapter 1

Core Principles for Instruction in Science

EFFECTIVE PEDAGOGY FOR SCIENCE CLASSES IN CHRISTIAN SCHOOLS BEGINS WITH a few core principles. Unpacking these principles and examining how they relate to classroom instruction will be the subject matter of the remainder of this book.

Within the broad spectrum of Christian schools in this country, some are part of the so called "classical and Christian" school movement. Those new to teaching in a classical school environment may find it difficult to articulate what makes science "classical." After all, the experimental methods we use today are radically different from the philosophical approach to nature found in the works of Aristotle. Science courses do not become classical just because they include readings from Aristotle or Ptolemy. It may also be difficult for those new to teaching in a Christian school to formulate effective ways of making science classes "Christian," or to articulate what it means to be a scientist who is a Christian, as distinguished from scientists who are not Christians. Science courses do not become Christian merely because they include Bible verses or Creationist teachings about macroevolution.

My own philosophy of education derives fundamentally from the first and greatest commandment, articulated by Jesus in Mark 12:30: "Love the Lord your God with all your heart and with all your soul and with all your mind and with all your strength." Loving God with one's mind, and teaching others to do the same, is the fundamental task of Christian education. Our minds must be developed so that we can serve God in his kingdom with all of the capability that our minds possess. To leave any part of one's mental faculties undeveloped would be to fail to love God with "all one's mind." This explains why Christian teachers and classical teachers love to read: They seek continually to develop their minds so that they can serve Jesus more effectively. This fundamental truth is also the primary connecting link between "Christian" education and "classical" education. Both are about developing the mind. Developing our minds so that we can use them for God's glory is the starting point for the core principles of Christian science instruction, which we will now proceed to develop.

Principle #1. The nature of scientific knowledge is just as important as the knowledge itself.

No scientific fact or principle can be fully understood or appreciated without also understanding the theoretical framework that enables us to interpret the fact or principle. This is because scientific knowledge, as we will discuss in later Chapters, is contingent. All scientific knowledge is theoretically based, that is, it must be interpreted and understood within some kind of theoretical framework. Theories, in turn, develop and change over time, so that our interpretations of the facts will change over time, sometimes radically. To gain a real understanding of the nature of specific scientific facts it is necessary also to appreciate the nature of scientific knowledge itself.

This situation is somewhat unique to the study of science. One can know a lot about American history without getting too much into what it means to know things. One can know a lot about Shakespeare without delving into the nature of the aesthetic experience itself, and what it means for human beings to be art producing creatures. But one cannot get very far at all in scientific study without understanding the contingent nature of scientific knowledge, and how our theories allow us to make sense of the facts.

In his 1922 Nobel Prize address Niels Bohr stated that scientists "believe the existence of atoms to be proved beyond doubt."[1] What could such a statement possibly mean in an era when atoms themselves were not yet well understood? The neutron wasn't even discovered until ten years later. Bohr's statement does make sense, but only if one acknowledges that scientific knowledge is contingent; it can change and often does. We now know, and have known for a long time, that atoms are nothing like the little planetary systems we sometimes imagine them to be. Scientific knowledge changes. Theories emerge, develop over an extended period of time, and are sometimes scrapped entirely. We may think we "know" a certain scientific fact. But really to know it, we need also to know what kind of critter a scientific fact is. (Pardon my metaphor – I do live and work in the capital city of Texas.)

If we are to be responsible science educators, and if our students are to be equipped with the best science education we can give them, then it is imperative that science teachers and students of science alike must understand the nature of scientific knowledge.

Principle #2. The history of science helps to illustrate how science works.

When it comes to realizing the first core principle, we are fortunate that the history of scientists and scientific discoveries affords a rich study in the way science works. Hundreds of well-known scientific discoveries, easily brought out in the course of primary and secondary science instruction, were made according to the ordinary process of scientific research, a process I call the Cycle of Scientific Enterprise (discussed in detail in Chapter 4). Reviewing these discoveries with students enables them to go way beyond ordinary discussions of the Scientific Method[2] that they

---

1 James Gleick, *Genius*, p. 38.
2 The so-called Scientific Method is introduced in every introductory science text. It is a useful concept, so I will retain the term. However, it should be pointed out that some scientists prefer to emphasize that there are actually many Scientific Methods.

start encountering in about fourth grade, and begin to become familiar with the broader mechanism by which scientific knowledge is amassed over time. Of course, it is appropriate for children to learn the basic steps of the Scientific Method, but those steps are simply a description of the way scientific experiments are typically performed. The Scientific Method does not explain the complex interaction between theory, hypothesis and experiment that drives scientific inquiry forward. At a higher level, knowledge progresses in major stages as the scientific community adopts various paradigms that guide research and determine the way theories and problems are framed. A familiarity with the ways these "paradigm shifts" have affected subsequent scientific inquiry is deeply enriching, and is very useful for connecting the principles learned in the classroom with the real world where scientific inquiry happens.

For example, a study of the Copernican Revolution, including major astronomers from the sixteenth and seventeenth centuries, and their discoveries and ideas, is one of the best ways to help students see how new theoretical models replace old ones, and how experimental evidence must be accounted for by a theory if the theory is to remain in favor in the scientific community. (This example is part of the discussion in Chapter 4.) Another example is the history of atomic theory. After John Dalton's brilliant first attempt at a detailed atomic theory, experiments conducted by J. J. Thomson, Ernest Rutherford and James Chadwick led to the formulation of new and more accurate models. As we will discuss in Chapters 3 and 4, scientific theories, which enable us to interpret scientific facts, are of paramount importance. When scientists are conducting research in new territory, each new discovery must be interpreted in light of a theoretical framework. The history of the development of the atomic model provides students with a clear sequence of easily understandable examples illustrating the interaction between experiment, theory and hypothesis.

Schools promoting a traditional liberal arts curriculum, and most Christian schools would subscribe to this description, strive to place a high premium on history in every discipline. Classical schools do this intentionally and overtly. But whether classical or not, any school that intends for its students to *understand* science, rather than merely memorize facts, should teach the history of the discipline in addition to the scientific content itself.

Principle #3. The tools of scientific thinking go much deeper than knowing scientific facts.

Imagine that you know a student who claims to have a thorough grasp of English literature. You begin asking the student what she has read and find that she has a ready knowledge of the story lines of every novel you ask her about. But then you begin asking her questions about themes, characterization, vocabulary, style and so on, and the student has no idea what you are talking about. This prompts you to tell the student that she may know lots of stories, but she doesn't really know literature at all. Literary study involves a lot more than just knowing plot, and a person who likes to read but who never digs into the deeper implications of the theme and structure of a novel can hardly be said to be a student of literature at all.

The same is true about studies in science. Knowing certain scientific facts, principles and laws is analogous to knowing the plot of a story. If this is all a student knows, then the student will not be equipped to "do science," that is, to engage in meaningful scientific inquiry or research. Yet, the overwhelming emphasis in most science programs is on content rather than on the tools that will enable students to explore the content. Preparing students for scientific thought and practice does, of course, require that they become knowledgeable of the facts and principles of current scientific models. One could hardly be considered to be scientifically literate without knowing, for example, Newton's Laws of Motion, the Law of Conservation of Energy, the basic structure of the Periodic Table of the Elements, and how these relate to the world around us.

But at a deeper and more important level, scientific thinking entails a host of cognitive skills that go beyond knowledge of specific scientific facts or laws. I call these "the tools of science," and I liken them to the tools a carpenter has in his toolbox. To make a piece of fine furniture the carpenter must have mastered the use of each of the tools, and he needs to keep the tools sharp and ready for use. Emphasizing the development of the tools of science in every science course at every level in the school must be one of the core principles embodied in any successful science program.

A basic list of the tools of science includes the following skills:

- The student can think of matter in scientific terms, observing, classifying and questioning such things as the chemical and physical properties of a substance, or the physical laws which govern physical and chemical processes.
- The student is knowledgeable of different units of measurement, including the USCS, SI and MKS systems of units, and is proficient at performing unit conversions.
- The student can describe the mathematical relationships embedded in the equations that express physical laws. Such relationships include direct and inverse proportion, independent and dependent variables, the behavior of different variables under different conditions, and linear and nonlinear functional relationships.
- The student can describe the steps in the Scientific Method, why they are necessary for a valid controlled experiment, and why this method has enabled science to be so successful over the past few hundred years.
- The student knows the difference between accuracy and precision and how both of these relate to taking measurements.
- The student can formulate a quantitative hypothesis.
- The student can effectively document an experiment in a lab journal.
- The student can write an effective lab report from scratch, including clear descriptions of background, procedure, and results.
- The student can use computer tools to create lab reports, including tables and graphs.
- The student can manipulate data, represent data in tables and graphs, and compute experimental error.
- The student can identify reasonable and significant sources of experimental error.
- The student can estimate the uncertainty in a measurement.

- The student is proficient at measurement techniques and standard laboratory practices, and is skilled at using proper care with laboratory equipment.
- The student has skills at exploring the uses and limitations of unfamiliar scientific equipment.
- The student can set up a scientific graph with appropriate labels and scales, and can use it to compare theoretical and experimental values and trends.
- The student can identify outlying data and suggest alternatives for dealing with outlying data.
- The student can confidently identify reasonable solution strategies for problems when a method is not readily apparent.
- The student can proficiently apply mathematics to the solution of problems.
- The student can use mental math to get quick, approximate answers.
- The student recognizes when calculated answers or experimental data are unreasonable or erroneous.
- The student appreciates the level of effort required to achieve mastery.

All of these skills should be addressed in an effective secondary science program. The list is long, and no single science course will address every one of these skills in depth. But most secondary courses should hit hard on many of the items in the list. Some of the skills (such as estimating uncertainty or exploring the limits of unfamiliar equipment) will typically be addressed only in upper level courses.

Principle #4. Teachers and students must approach scientific study with a distinctly biblical worldview or philosophy.

It may seem like stating the obvious that the science program at a Christian school should reflect a biblical worldview. However, realizing this principle requires a tremendous degree of sophistication on the part of the science teachers, and is not at all as straightforward as it sounds. Every Christian teacher affirms and every Christian school promotes the truths that God created all there is, that his handiwork—the created order all around us— is wonderful, and that the created order exists for his glory. But beyond constantly conveying these truths to our students, a biblical approach to science also entails an emphasis on the following propositions:

- *Surface treatments such as inserting Bible verses into textbooks don't do the job.*

    Nor does aligning the curriculum with one narrow perspective on the age of the earth or evolution. Christianity is a big tent, and views on these matters vary widely among Christians. We will have more to say on this controversial topic later. The point here is that making science studies biblical is really about loving God with one's mind – scientist, teacher and student alike. This is the case for all academic disciplines. We seek, in any discipline of study, to understand the world God made, to be salt and light, to seek justice and mercy, to acknowledge the reign of Christ, and to help extend the boundaries of his kingdom as we are given opportunity.

- *Biblical faith informs research, and scientific research informs faith.*

    By saying this I am not implying that the Bible should be used as a scientific text. What I mean is that biblical faith both encourages scientific inquiry and points research in a productive direction. If Christianity is true, how could it be otherwise? Central to Christianity are the claims that Jesus Christ is the Lord of the Universe and that as his people we should pursue our work to his glory. Just as in any other field of endeavor, the faith of Christians pursuing scientific research will have a bearing on their work. For example, I believe that it is God's intention for us to live on this planet without contaminating it with pollution. Consequently, I believe there are abundant sources of energy available to us that do not pollute. If I were a researcher investigating this issue my research would be fundamentally driven by this belief, which stems from my faith in God's intention regarding how we should live. An unbeliever investigating the same question might hope that there were practical, alternative sources of energy, but could have no internal certainty that they were there waiting to be discovered.

Conversely, scientific research can and does inform Christian faith and Christian doctrine. This was exactly what happened when the Copernican Revolution eventually resulted in the Church's acceptance of the heliocentric model for the solar system, and the subsequent revision of doctrine concerning man's place in the created order. Prior to the 16th century Church doctrine held that Psalm 19:4-6 affirmed the orbit of the sun around the earth. This passage, which states of the sun, "Its rising is from one end of the heavens, and its circuit to the end of them," must be interpreted in some other way now that it is known that the sun, in fact, does not orbit the earth.

The methods and practices used by a responsible scientist in his or her work do not depend on whether the scientist is a Christian or not. All good scientists will use the utmost care to develop sound theories and conduct valid experiments that will reveal unambiguous data and useful interpretations of them. But a believing scientist knows that the things he or she discovers about the natural world are the way they are because of God's wonderful and inscrutable purpose, while the unbeliever may approach his or her studies without theological considerations.

- *Science and faith complement one another; when the claims of each are rightly interpreted they do not conflict.*

Over the past few hundred years many different views have been proposed regarding the relationship between science and faith.[3] Most of these views must be regarded as unacceptable from a biblical point of view. One such view is that faith and science simply view the world from different perspectives that do not relate to one another. Science gives us the facts of nature, and faith gives us the meaning of those facts.

---

3 The outline of four views that follows is from an interview between Kenneth Myers and Stanton L. Jones in the *Mars Hill Audio Journal*, Vol. 50, May/June 2001.

This view seems helpful on the surface. However, the Bible gives us too many instances of the spiritual realm intersecting with the natural realm for us to keep these apart. The burning bush either happened or it didn't, and Jesus was either an ordinary human being or he wasn't. A philosophy that holds that Christianity only deals with the spiritual world without ever intersecting the physical will inevitably call into question the miracles attested to in the Scriptures and ultimately faith itself.

Another obviously unacceptable view, popular these days among the community of atheistic scientists who believe that everyone else needs to be an atheist too, is that religious faith of any kind is fatuous, maybe even destructive, and that science, as it explains things with increasing power, will ultimately leave no room for God. Eventually, all vestiges of religion will be eliminated from our culture as people learn how absurd religious belief is. Clearly, no Christian holds such a view, but many contemporary commentators do.

A third unacceptable view became popular with the rise of Post-modernism in the late twentieth century. This view holds that science is just one competing "metanarrative" among many and holds no special authority. Its truth claims are relative and motivated by the desire for power, as are the truth claims of every other political group. Now, I believe that Post-modernism must be credited with helping us to be aware that we all do bring our own perspectives to knowledge, and that strictly speaking, there is no such thing as a purely objective observer or reporter. However, extending this idea to the extreme of denying the validity of scientific knowledge, and further to the denial of all absolutes whatever, is clearly problematic for Christians who do believe in absolute truth.

A fourth view is one that I find healthy and productive. This view, sometimes known as Critical Realism, recognizes that God has made a real world, and has equipped humans to engage and know the world, and to know it truly, although not exhaustively. We have a legitimate capability to learn things about the world we live in, to communicate these to each other, and to use this knowledge to be better inhabitants and stewards of the earth. We do not know and probably cannot know everything there is to know about the natural order, so there will inevitably be gaps in our knowledge that appear to put faith and science into conflict. Also, we know we are prone to error, which keeps us humble, and we take this into account when we make claims regarding truth or scientific facts. But we affirm that the claims from the realms of faith and science, when rightly interpreted, do not conflict.

- *Science provides us with an excellent forum in which to love our neighbor, and in which to learn what loving our neighbor means.*

To the extent that scientific discoveries benefit humanity, we are loving our neighbor while we are doing science. Of course, scientific research is often impersonal, but it is not always so. Many scientists have been inspired to research in certain fields because of the suffering they have witnessed on the part of patients, friends or family members. But impersonal or not, God has commanded us to love our neighbor, and research that is directed at benefiting humanity is an act of love.

But the enterprise of scientific research is also a tremendous crucible that confronts us with a thousand ethical questions associated with the proper ways of pursuing research and applying the results of it. We must explore, and teach our students to explore, ethical issues associated with conducting research such as humane treatment and research methods, faithful and honest reporting of results, responsible use of resources, respecting dignity and privacy, and a host of other issues. We must also teach our students to explore ethical ways of applying the results of scientific research in contemporary communities and markets. The results of scientific research are often misused, sometimes grossly, adding to environmental contamination or contributing to injustice. Instances of this in the twentieth and twenty-first centuries abound, and have many times been egregious or plainly heinous. A biblical approach to the discipline of science entails learning from these errors and teaching our students to honor Jesus Christ by the integrity with which they approach such challenges.

- *If we are to serve Christ faithfully in the discipline of science we must be responsible learners.*

    In the mid-1980s I attended a lecture at the University of Houston given by the late paleontologist Stephen J. Gould. As I recall, the lecture was on the subject of lessons we can learn from the John Scopes trial of 1925, the first highly publicized legal clash over the teaching of evolution in the classroom. Professor Gould began his remarks with an extended criticism of Christians who, in their efforts to fight evolution, had misrepresented him, misquoted him, spelled his name wrong, misrepresented his academic credentials, and so on. In his slides he showed many examples of writings by Christian authors containing these kinds of errors. As a Christian examining these documents I was appalled and saddened at how my brothers in the faith had done such disservice to our cause by producing such shoddy work. Professor Gould was right to be upset about this, and I am most thankful that I was there to see it, because it left a lasting impression on me. If Christians wish for their views to be respected when shared in the public realm, then they need to exercise the same degree of care in their research, citations and arguments that they expect of others.

    In our endeavor to be responsible learners we must also be respectful toward our opponents. Anyone who has ever attended a university lecture on the Intelligent Design—Evolution debate knows that the climate that prevails in such gatherings is often anything but respectful. Mockery, yelling, *ad hominem* irrelevancies and outright insults are common from people on both sides of the controversy. Sadly, it is common to find people gleefully relishing every instance they can find of factual error or inconsistency or hesitancy when answering questions. Instead of a respectful dialog in which parties from both sides seek correctly to understand the scientific data, graciously overlooking or gently correcting faults just as we would wish others to do for us, we have all contributed to the prevailing climate of hostility. As another example, it is irresponsible to make statements such as "Darwin was an evil man," as some Christians have. Aside from what any of us may think about the theory of evolution, Charles Darwin was a consummate scientist, a brilliant researcher, and a bold theorist. No one who has read *On the Origin of Species* (as all Christian science teachers should do) can deny this. Darwin was no more nor less evil than the rest of us, and calling him so is poor teaching. If our students are to learn a biblical way of approaching science, they must learn from us how to hold respectful discourse with those with whom we disagree.

Finally, responsible learners are open to see what the data have to show us, even if scientific findings seem to conflict with the views we hold. Many Christians seem to have their minds made up that global warming, if it is happening at all, is a natural cycle and is not being caused by human activity. Some are convinced that the earth is only a few thousand years old. Some Christians hold that the dinosaurs are a hoax and did not exist, or possibly that they co-existed with man. A large number of Evangelicals hold that there is no way that God could have used macroevolution as part of his creative work. And some are apparently uninterested in environmental issues, holding that Scripture's "subdue the earth" mandate authorizes humans to use the earth and its resources at will. It is not my intention to make pronouncements on any of these issues. But I submit that since new data are constantly coming to our attention we can only come across as hypocrites to our students if we refuse to review new evidence for our opponents' views while insisting that they accept our views. Instead, we must promote by both word and example the true ethos of responsible, careful inquiry and evaluation.

Chapter 2

Truth and Facts

Because of the abundance of references to truth in Scripture, truth is precious to Christians. Many of these references are well-known: "I am the way, and the truth, and the life" (John 14:6), "You will know the truth, and the truth will set you free" (John 8:32), "God is spirit, and those who worship him must worship in spirit and truth" (John 4:24), and many others. We are all aware that the notion of absolute truth has come under fierce attack during the last few decades. This attack is especially challenging for Christians because we hold the fundamental truths of our faith to be absolute and unchanging, even though doctrines do develop over time. The truths "God made all things," and "Jesus is Lord" are true for all time, whether one accepts them as truths or not. They are true for all people, in all places, at all times and can never change. A correct understanding of the nature of truth is critical to us because our faith is grounded in what we believe to be the truth about God, Man, and the rest of Creation.

As Christians engaged in teaching and learning topics in science, our affirmation of things that are true loses force and clarity if we do not have solid definitions to distinguish between propositional knowledge we would call truth, and other types of knowledge. We hold truth to be absolute and unchanging. However, "scientific facts" (which I will refer to simply as facts) are contingent; they can and sometimes do change as new data and information come to light. For example, prior to 1500, astronomers maintained as fact the belief that there were seven heavenly bodies. Additional facts from this period were that these heavenly bodies all orbited the earth, and that they did so once every 24 hours. However, during the Copernican Revolution it became increasingly clear that the planets were orbiting the sun, not the earth. Further, Galileo's telescope revealed that there were more than seven heavenly bodies. So the facts about our system of planets began to change. The best we can say about scientific facts is that they are supported by a lot of experimental evidence (data), and they are correct as far as we know.

Truth is central to Christianity, and as Christian teachers we wish for our students to appreciate the permanence of the truths in which we believe. But if we confuse the terms "truth" and "fact" in our teaching by using them synonymously, our students might be misled into thinking that because facts change truth changes, too.

A deep exploration of the distinction between these terms rapidly takes one into the realm of epistemology, a taxing and complex subject. But this subject is so important, and so fundamental for how science is taught in the classroom, that we will have to go far enough into this realm to form a proper distinction between truth and facts. Further, students are bound eventually to raise questions about different types of facts, so we will need to address that as well.

Let us begin with some working definitions.

Truth: A proposition that is true for all time, all places and all people. Truths never change. We know truths by revelation or first hand testimony.

Fact: A proposition that is supported by substantial experimental or observational evidence (data), and which is correct as far as we know. Facts can change as new data and information become known. We know facts by observation and experiment, or by making inferences from our observations and experimental results.

I maintain that in the context of science classes the term "truth" should be reserved for propositions made known to us by revelation. Christian theologians generally hold that the revelation of God to man is in two "books": the book of God's Word, the Bible, and the book of God's Works, the creation. The revelation found in the Bible is sometimes referred to as "special revelation," while revelation from the creation is referred to as "general revelation." Of course, most Christians are completely comfortable with the notion that the Bible contains revelation from God. This is, in fact, the internal claim of the Scriptures, affirmed over and over. The classic reference for this claim is 2 Timothy 3:16-17: "All Scripture is breathed out by God and profitable for teaching, for reproof, for correction and for training in righteousness, that the man of God may be competent, equipped for every good work." Moreover, the Old Testament prophets repeatedly claim their revelations to be from God. Finally, John begins his Revelation with this claim: "The revelation of Jesus Christ, which God gave him to show to his servants the things that must soon take place."

In addition to the special revelation of the Scriptures, the Bible itself informs us that the creation also reveals truth to us. Romans 1:18-20 states:

> *For the wrath of God is revealed from heaven against all ungodliness and unrighteousness of men, who by their unrighteousness suppress the truth. For what can be known about God is plain to them, because God has shown it to them. For his invisible attributes, namely, his eternal power and divine nature, have been clearly perceived, ever since the creation of the world, in the things that have been made. So they are without excuse.*

And a classic and beautiful passage about general revelation is the first section of Psalm 19:

> *The heavens declare the glory of God, and the sky above proclaims his handiwork.*
>
> *Day to day pours out speech, and night to night reveals knowledge.*
>
> *There is no speech, nor are there words, whose voice is not heard.*

Their measuring line goes out through all the earth, and their words to the end of the world.

In addition to revelation, there is another very common and quite legitimate use of the term "truth." This is in instances involving first hand testimony. In our courts we require witnesses to swear that they will tell the truth. A parent questioning a child about his behavior exhorts him to tell the truth. We want our politicians to tell us the truth. Each of these cases involves first hand testimony. The person making the statement is making a truth claim about his or her own experience. Whether the claim is credible depends, of course, on the character of the witness. Although common in everyday human relations, and important when we trust scientists to tell us the truth about their findings, this use of the term truth does not arise in any particular way in the study of science in school classrooms, so I will not address it any further here.

The following tables list a few examples of propositions — truths — that have been revealed to us, either in Scripture, or in nature, or both. I would like to note that coming up with good examples of truths revealed in nature that are not also revealed in Scripture is quite challenging, because the Bible is so exhaustive in its teaching. I have made an attempt in the second table below to identify a few such truths, but you will notice that all but the first are descriptions of human nature.

Truth revealed in Scripture, but not obvious in nature (Special Revelation)	
1	God is love.
2	There is one mediator between God and man, the man Jesus Christ.
3	Hatred is evil, and is morally equivalent to murder.
4	Human sexuality and marriage mirror the relationship between Christ and the Church.
5	Jesus is the atoning sacrifice for our sins.
6	We are saved by grace through faith.

Truth revealed in nature, not obvious in Scripture (General Revelation)

1	God is a master of complex mathematical patterns. (Simple mathematical patterns appear in Scripture, such as the recurrent use of the numbers 3, 7, 12 and 40. But there is nothing in Scripture like $\left(\dfrac{T_1}{T_2}\right)^2 = \left(\dfrac{R_1}{R_2}\right)^3$ or $F = G\dfrac{m_1 m_2}{r^2}$, which describe the natural order.[4])
2	Personal experience of a tragedy gives one knowledge of suffering that cannot be obtained any other way. (Prov. 14:10 hints at this.)
3	Most human parents have an intense love for their children that does not extend equally to other children.
4	Typically, mastery of any human endeavor requires sustained practice and discipline, and will fade when these cease.
5	In most cases, if a man is imprisoned and treated in a severe and inhumane fashion his spirit will be sorely tried and will eventually, if the treatment continues long enough, be broken.
6	Romantic love can lead people to act irrationally.
7	In most cases, people can be induced through torture to "confess" to acts they did not commit.

Truth evident both in Scripture and in nature (Special and General Revelation)

1	A divine being, God, exists and made all things.
2	God is eternally powerful.
3	All men are sinful.
4	Men and women, while very similar, are different in key respects and play overlapping but distinct roles in community, society, and family.
5	The creation is loaded with beauty, of which God is the source.
6	We each must experience physical death.

In contrast to truth, scientific facts are discovered by observation and experiment, or by making inferences from these. When

---

4 That such relationships exist in nature is properly regarded as a scientific fact, evident to every scientist, believer and nonbeliever alike. The truth behind this fact, for those who have eyes to see it, is that this mathematical order came from God.

using the term "fact" we should be aware that scientists tend to use this term only when speaking to lay audiences. When speaking to one another they simply refer to the data and the inferences drawn immediately from the data, which is what they mean by the term "fact." General inductions about the nature of the physical world drawn from these data are called "theories," a term we will explore in detail in Chapter 4.

The table below lists a few examples of well-known scientific facts.

Scientific Facts, all correct as far as we know	
1	Sodium atoms, when excited, emit orange light at a wavelength of about 590 nm.
2	There are approximately 100 billion stars in our galaxy.
3	The stars and galaxies are all rushing away from each other. That is, the universe is expanding.
4	Our solar system, with its eight planets, is heliocentric.
5	Hydrogen atoms in the sun fuse to form helium.
6	Most tropical fruit trees reproduce asexually by a process known as agamospermy.
7	The average atomic mass of carbon is 12.011 grams/mole.
8	The instructions encoded in the genes in the DNA of an organism are responsible for the physical characteristics expressed in that organism.[5]

Looking more closely at scientific facts, we should distinguish between those facts discovered by experiment and observation (or inferences from these), and those which are correct by definition. The facts listed in the table were all discovered by experiment or observation, or by drawing inferences from these. They are correct as far as we know, but in principle they could change as new information becomes known. But there is another kind of scientific fact, facts that are correct by definition. For example, it is a scientific fact that salt dissolves in water, but axle grease does not. This is correct by the definition of "dissolving," and according

---

[5] This fact is a good example of one that is probably about to change. The emerging science of epigenetics indicates that organisms can pass on acquired traits after all, by the epigenetic mechanisms that switch genes on and off.

to the definition of the term, cannot become incorrect in the future. The term "dissolve" was coined to describe something that salt does in water but axle grease does not do. A slightly different example of this came to our attention in 2006 when the official criteria for designating an object in our solar system as a planet changed. Until that year our solar system had nine planets, but when the criteria for planethood changed Pluto was ruled no longer to be one. In this case the fact (correct by definition) of how many planets there are did change because the definition of what a planet was changed.

A word should also be said about the difference between historical facts and scientific facts. While an historical event is occurring, historical facts function similarly to scientific facts. For example, we can say that Kofi Annan was Secretary-General of the United Nations from 1997-2006. We know this because we observed, as they happened, the reports in the news, and we remember them. However, in the future, when everyone now living has died, the only way this fact will be known will be through historical records. Unlike scientific facts, which can be verified any time by experiment or new observations, historical facts cannot be experimentally repeated for verification.[6]

Cautions and Qualifications

The main point of this Chapter has been to distinguish between contingent knowledge like scientific facts, which can and do change, and permanent truths. However, as I said earlier, epistemology is a complex subject, and with any attempt to provide a simple organizational structure for a complex subject there will be limitations. It is important that we recognize the limitations of my approach to dealing with truth and facts.

First, you may have already noticed that the categories actually overlap. One of the ways we know facts is to make observations of the physical world. But general revelation is also based on observations of the physical world, as the passage in Romans 1

---

[6] There is an interesting overlap between scientific facts and historical facts in the astronomical observations of past events. These obviously cannot be repeated, and we must rely on historiographical methods for verifying those data.

states. What is the difference between the observation that the sun rises each morning and the observation that humans are often given to lying? Or what about the resurrection of Jesus? Is it an historical fact, as reported by the testimony of eyewitnesses, or is it part of special revelation, taught to us in the Scriptures? Or both? I frankly admit that the categories I have set up (truth and fact) cannot be rigidly maintained when scrutinized at this level.

What we are really talking about here is that there are different kinds and sources of knowledge. I believe the distinction I have developed between truth and facts is useful, particularly in primary and secondary classrooms in which students are too young to have explored the nature of knowledge more extensively. It is particularly useful in light of how scientific knowledge is developed over time through theory, hypothesis and experiment, always improving in accuracy but never arriving at completeness (as discussed in the next two Chapters). It is important for our students to understand that there is a difference between our understanding of how atoms are constituted, which may change radically as new experiments are developed, and what we know about how we obtain salvation, which Scripture clearly states is through Jesus Christ and no one else.

So this Chapter presents a particular model of how knowledge works. Interestingly, in the next Chapter we will be talking about how we use models in the construction of scientific knowledge in particular. But I think this simple model of two different types of knowledge is useful for students in the primary and secondary levels. As they mature and consider these questions more deeply this model will need to be improved and enhanced with more complex categories, a more detailed description of the interactions between the categories, and so on. This is what epistemology is all about.

The usefulness of the two categories I have set up (truth and facts) is significantly increased if we keep in mind that a "fact," to a scientist, is a proposition that is very close to the data. The truths we know from General Revelation, while based on observation, are somewhat removed from the immediacy of the data. Thus, in many cases, we can maintain the distinction between truth and facts afforded by my definitions, even when both are based on observations. For example, the Bible's statement that

God's "eternal power and divine nature" are clearly perceived from what has been made is obviously about observations of nature, but the truth at issue is several levels removed from the raw data of sunrises, starfish, and skeletons. When we observe these data, and contemplate them, we eventually arrive at the truth that there is a powerful divine Being responsible for them, but this truth, as I've said, is several removes from the raw facts.

Second, in my description of the difference between truth and facts I have as yet made no reference to the levels of certainty involved with either. This is complicated too, and should be discussed with students as soon as they are ready, which will probably be indicated by the questions they ask. Just because we believe specific truths as part of our faith does not mean we possess absolute certainty about them. In fact, it is often the case (I know it is with me) that the more deeply we think about our doctrine, and the more wise people we meet who believe differently than we do about things, the less willing we are to declare that we possess absolute certainty.

I have not really been able to see any end to this conundrum. At the level of denominational differences we are probably all willing to admit that on many issues we cannot say for sure who is right. So our certainty about things is something obviously less than absolute. I attend a church that holds to certain views about baptism, church governance, and church membership, and many other churches hold different views. Are any of us ready to declare absolute certainty about who is right on such questions? I doubt it.

If we go down to the level of, say, the Apostles' Creed and ask if we have absolute certainty about this description of our beliefs, I for one would be much more confident but I would have to confess that I still could not claim to have absolute certainty, even about the Apostles' Creed. Going down to an even more basic level, I do know (from the historical record in Scripture) that Jesus' words to the thief on the cross imply that whatever the minimum requirements were for salvation, the thief met them. The irreducible minimum is apparently a simple faith in Jesus.

What about at this level? Am I *certain* that a simple faith in Jesus will result in eternal life? Well, it is emphatically certain that I believe it, but am I *certain* that I am right? I am very certain that Christianity teaches this, and it seems clear in the Gospels that Jesus made this claim. But as we all know, most of us go through times of doubt, sometimes even storms of doubt. So even at this most basic level, the issue of certainty is a thorny one that is bound up in the very psychology and frailty of our humanity. We can't escape this. We are fallen creatures who "see in a mirror dimly" and who "know in part." But we are also told that one day we "shall know fully, even as we are fully known." It seems to me that for absolute certainty we will have to wait until that day. It should surely keep us humble to discover that we sometimes feel more certain about some scientific facts (such as that a year is approximately 365.25 days) than we are about the truths of our faith.

Third, I have not referred to the way revealed truths have been unfolded to us through time. A devout Hebrew in Old Testament times would have probably been scandalized by the New Testament teachings about the Trinitarian God. This is, in fact, one of the factors that led to Jesus' death, as John 10:33 plainly describes. We could say the same thing about the progressive revelation associated with monogamy, or taking of life, or a number of other doctrines. So revealed truth itself seems somewhat of a moving target.

I don't think any of these factors detract from the usefulness of the distinction between truth and facts that I have set forth. However, we must recognize that this model is a simplification of the way knowledge works, and we must exercise caution and maintain humility regarding our own claims to certainty.

Application to the Classroom

The distinction between truth and facts, using the simple definitions reviewed here, should be introduced to students in the third grade, and reinforced each year after that. Students arriving in ninth grade should be able to respond with a few well-written sentences to this prompt:

> Susan is talking to her friend Jane over coffee. Jane has a Ph.D. in astrophysics. Susan says, "Jane, I am telling you the truth: Jesus Christ is the only way to God." Jane replies, "To me, truth is things like the universe starting with the Big Bang. I *know* that is true. This stuff about Jesus, I don't know if that is true or not."

> Using what we have discussed in class, briefly describe how the word truth is being used in this conversation. If you were Susan, what would you say next?

A good response to this item would begin by noting that Jane's comment about the Big Bang should more properly be regarded as a scientific fact (correct as far as we know), or even better, as a theory (a generalization from the observed fact that the universe is expanding), but not as a truth. The student would then elaborate on the difference between truth and facts, and how this distinction could be used by Susan to further her discussion. Here is another example:

> One of my college professors once said, "You people need to get over it. Evolution is a scientific fact!" During the break one of the students in the class said, "I just can't believe what our prof said—There's no way that evolution is true."
>
> Comment on these statements using what we learned about the nature of scientific knowledge.

An appropriate response to this prompt should certainly focus on the student's statement in which the term "true" was substituted for the professor's term "fact." Students should clarify the distinction. Students may also address the difference between facts and theories (the topic of Chapter 4), and suggest that the professor may have been more correct to stick to the term theory when describing evolution.

Students in higher grades can be expected to discuss the nature of scientific knowledge in more depth, including the recognition that some facts are correct by definition, and the basic difference between scientific facts and historical facts.

Chapter 3

What Is Science?

WHEN I POSE THIS QUESTION TO MY NINTH GRADE SCIENCE STUDENTS I OFTEN get answers that emphasize learning about the natural world. Certainly the field of exploration open to science is the natural world, but science is a much larger endeavor than just amassing scientific facts about nature. Scientific facts are nearly meaningless without some means of interpreting them and relating them to other facts. How far would medicine have progressed if Alexander Fleming had not connected the fact of human susceptibility to infection to the fact that a blue-green mold killed *Staphylococcus* bacteria? In isolation, the effect of the mold on the bacteria is interesting (at least to someone who is fascinated by mold). But interpreted together with other facts in light of an emerging model of how the human body responds to bacteria the discovery, which led to the commercial development of penicillin, was revolutionary.

The key word in this example, and in my definition of science, is "model." Here is my definition for science:

Science: Science is the use of experiment and observation to develop mental models of the natural world. The models we develop are called theories and are for the purpose of accurately describing the properties, ordinary functioning, and history of the natural world.

A model is more than a single piece of information. Knowing that battleships are gray does not mean that "gray" is a model of a battleship. Models are complex structures, incorporating many details, which represent some other object or system. For example, in the repair manual for my automobile is a wiring diagram for the car's electrical system. This diagram is a model (on paper) of the electrical system. It is a representation of the system (hopefully an accurate one) developed for the purpose of servicing the vehicle.

All models share certain characteristics. First, a model is designed to be detailed enough to serve the purpose for which it was developed. Second, there is a limit to the amount of information (detail) that can be included in any specific part of the model. Third, all models are incomplete, that is, no matter how detailed they are all models fall short of the reality they represent. In order for a model to be a complete representation of reality it would have to be, or even transcend, the reality itself, and a model used to describe a reality cannot *be* that reality.

Applying these three characteristics to my wiring diagram example, we first see that the wiring diagram needs to be detailed enough to allow a mechanic to service the car. The diagram needs correctly to identify every wire in the system, where it originates, and where it terminates. The diagram would be even more helpful if it indicated the colors of the wires, since otherwise it would be difficult to determine which one was which when servicing the vehicle. But if the diagram omitted various bundles of wires because the person drawing the diagram didn't know about them, then we would have to say that the model could be improved by being more complete. Second, it would be superfluous for this model to include, say, information about the compounds used to make the insulation on the wires, or the exact formulas for the pigments used in the coloring in the insulation of the wires. This information is important for someone, but not for auto mechanics. If the publisher of the repair manual began including such information, the repair manual would contain hundreds of volumes of information that mechanics would never use. Finally, no matter how much information was included in the wiring diagram of my car's shop manual, the wiring diagram would never be a complete representation of the actual wiring. The only exhaustively complete representation of the wiring in the car is the car itself with its wiring. But the car is not a model, it is itself. And if the purpose of developing the wiring diagram was to describe the wiring so that it could be serviced, then the car itself is not useful as a model.

Let us now apply these three characteristics of models to the field of science. First, our mental models of the world need to be detailed enough. Our present understanding of the kinetic theory of gases, of the patterns in the Periodic Table of the Elements, and of mammalian biology certainly is very detailed. Theories associated with these domains of scientific knowledge serve us very well, have enabled vast areas of scientific research to proceed successfully, and have given rise to an ocean of new technologies. On the other hand, neurologists would say that our present models of how the human brain works are still quite limited. We know a lot more than we used to, and we have recently seen some fascinating new technologies based on our current model of brain functionality, but everyone knows that research is just getting started in this field. Decades of research remain before our understanding of the human brain can be called thorough.

Next, there is a limit to the amount of information that can be presented in any part of a model. This is precisely why we divide science up into different fields (e.g., low temperature physics, cell biology, biochemistry). We humans simply don't have the mental capacity to consider everything at once, so we divide things up and discuss them in separate pieces. Of course, these pieces all relate together, so that the deeper one goes in studying science, the more one has to know in order to understand new concepts. One may not need to know much of the history of Nepal in order to have a solid grasp of the history of Texas. But one must know a great deal of physics, chemistry and biology to understand anything about the workings of a human blood cell. There are serious implications of this for science pedagogy. Science, like foreign languages and mathematics, is a cumulative discipline. Whether or not the History Department requires students to remember their ninth grade history during all four years of high school is a very complex question and is for the history faculty to decide. But the teacher of the senior Biology II class depends on his students remembering their physics and chemistry. If they don't, they simply cannot proceed in their studies of biology. We will address this in more depth later.

Finally, all models are limited and fall short of the reality they represent. Further, this will always be true because our knowledge of the physical world will always be less than total. The litmus test for judging this claim is to ask if it is possible for humans to apprehend the natural world in the same way that God apprehends the natural world. The obvious answer to this question (at least it seems obvious to me) is that since humans are limited in their intellectual powers and God is not, it is not possible for humans to know the world as exhaustively as God knows it. Thus, our mental models will always be incomplete and there will always be more scientific research to do.

Not all scientists agree with this claim. In his 1998 book *Consilience*, atheist Edward O. Wilson affirms his belief that human beings have the capability of exhaustively comprehending the universe (although why we should have developed such powers accidentally by evolution is a perplexing problem for him). However, it is my contention that a proper humility in the face of our own limitations, as Scripture urges upon us, compels us to acknowledge that our models will never be complete and are thus always subject to expansion, revision, improvement or replacement as long as this world, and human scientific research into its nature, continues.

The way scientists develop the mental models we call theories can be illustrated by this example from a ninth grade physical science course. After learning about the conservation of energy and the ways energy transforms, students can consider the energy from burning fuel that is used to heat water in a steam boiler. When asked where the energy from the fuel went, all of the students will reply that it went into the water and the steam. But if pressed further in a class discussion about where exactly this energy is, and in what form, the students might slowly begin piecing together an explanatory model. At some point a student will suggest that the energy is in the individual water molecules. This wonderful suggestion should prompt the next question, which is what form the energy takes in those molecules. Eventually students can figure out that the hotter molecules will be moving more rapidly.

That is, they possess more kinetic energy. This will prompt the definition of the internal energy of a substance. Further discussion can lead to the insight that if the kinetic energy of these molecules becomes great enough they will actually leap out of the liquid phase and into the gaseous or vapor phase to produce the steam in the boiler.

This train of thought can be pressed yet further (perhaps in a second day of discussion) until students can explain the pressure inside a container in terms of the molecules inside colliding with the container walls, and that hotter molecules, which move faster, will collide more forcefully and produce a higher pressure inside. A great example for helping students to see this is what happens to one of those helium-filled Mylar birthday balloons one buys at the grocery store. The balloon will seem normal inside the store. When taken outside in the summer heat and sunshine the balloon will swell up so tightly it appears ready to burst. But then when placed inside a car with the air conditioning system running at its maximum, the air in the car gets quite cold and the balloon will actually shrivel up and compress to the extent that it doesn't even float in air any more. It will sink to the floor inside the car. At the end of this extended discussion the student will have developed a mental model of the kinetic theory of gases, just as the scientists in the nineteenth century did when they were figuring all of this out!

Application to the Classroom

The notion that science is akin to model building needs to be introduced around fourth grade. Students need to understand that science is not a static discipline (if there is such a thing) requiring one merely to learn the facts. Science is a way of approaching and acquiring knowledge, a certain kind of knowledge, of the natural world. In many science programs the Scientific Method is introduced in fourth grade. This is an appropriate context for initial explorations into the nature of science and how scientific knowledge progresses. This subject will be addressed in much more detail in the next Chapter.

Chapter 4

The Cycle of Scientific Enterprise

THIS IS AN IMPORTANT CHAPTER. A MAJOR CONTRIBUTOR TO THE UNFORTUNATE clash between many Christians and the claims made by the scientific community is the failure to understand what theories are and the role they play in what I call the Cycle of Scientific Enterprise. I hasten to add that Christians are not alone in misunderstanding the nature of scientific inquiry; this problem is widespread in our country and represents a general failure of science pedagogy in schools over many decades. But since evangelical Christians are very vocal about issues on which prevailing scientific opinion appears to conflict with the Bible, it is crucial that we develop a proper understanding of how scientific knowledge progresses. Key concepts embedded in the Cycle of Scientific Enterprise are scientific facts, theories, hypotheses, and experiments, and the roles each of these play in the ongoing progression of scientific knowledge.

Here is a simple test to illustrate my claim that many people do not understand the key terms associated with scientific inquiry. Imagine that your car won't start one morning. Instead, you hear only a clicking sound and the engine doesn't turn over.

Imagine further that you know your battery is six years old, so you suspect that the problem is a dead battery. Is the following statement appropriate? "My car wouldn't start today. My battery is pretty old, so my theory is that the battery is dead and if I replace it things should be fine."

The short answer is that this is not an appropriate statement to make. Again, imagine that a friend at work hears that your car wouldn't start and comes over to commiserate with you about it. Consider these three statements that you might make:

Well, my hunch is that my battery is dead.

Well, my hypothesis is that my battery is dead.

Well, my theory is that my battery is dead.

Most people would probably speak the first or the third of these statements, but from a scientific point of view the middle one is correct. The first statement might be fine if you weren't speaking scientifically at all, or if your reasons for suspecting the battery were more nebulous. But since you are basing your speculations on data (the sound you heard and the age of your battery), your guess about the cause of the problem is more scientific than the word hunch connotes. A scientific hypothesis is a proposition about a scientific problem, based on a particular theoretical perspective. Hypotheses can be tested by experiment, which is of course true in the case of the battery since one can easily replace the battery and see if the car will start.

Referring to your speculation as a theory would not be appropriate at all, as I will discuss below. A theory is not a guess, it is not a hunch, and it is not an hypothesis. Understanding what theories are and how they function in scientific inquiry is at the heart of the issue addressed in this Chapter.

Let us begin with some working definitions:

Fact: A proposition that is supported by substantial experimental or observational evidence (data), and which is correct as far as we know. Facts can change as new data and information become known. We know facts by observation and experiment, or inferences from these. (The nature of scientific facts as knowledge was explored in Chapter 2.)

Theory: An explanatory framework or interpretive perspective that allows us to make sense of the facts; a "mental model" of how the natural world works in a particular arena of scientific exploration. Theories are the primary, central pillars supporting the entire structure of scientific knowledge.

Hypothesis: A specific, positively stated proposition, based on and derived from a theoretical framework, which can be tested by experiment. Hypotheses may often be stated in IF-THEN form, such as, "If the gas is heated at a constant pressure, then its volume will increase in direct proportion to its temperature."

Experiment: A controlled test of an hypothesis specifically designed to isolate, test, and measure the relationship between two or more variables in a physical system by manipulating one and measuring the effect on the others.

I should pause at this point to note that the Cycle of Scientific Enterprise, which we are about to discuss in detail, together with the definitions above, are themselves part of a model, a model of how science works. It is a model designed to help students understand how science works most of the time for most people. But as with all models, it has its limitations and necessarily incorporates simplification.

The figure on page 35 illustrates how the four entities defined above relate together in the Cycle of Scientific Enterprise.

The Cycle of Scientific Enterprise is a more comprehensive attempt to describe how science progresses than the standard six-step process we call the Scientific Method. The Scientific Method focuses mainly on the process of conducting a valid experiment, an important step in the development of scientific knowledge. Of course, the first step of the Scientific Method ("state the problem") clearly relates to a situation in which a theory does not provide an adequate explanation for some phenomenon, so the Scientific Method does include this one connection to the elements in the Cycle of Scientific Enterprise. But, as will be explained below, the Cycle of Scientific Enterprise also describes how experimental results relate back to the ongoing formation of

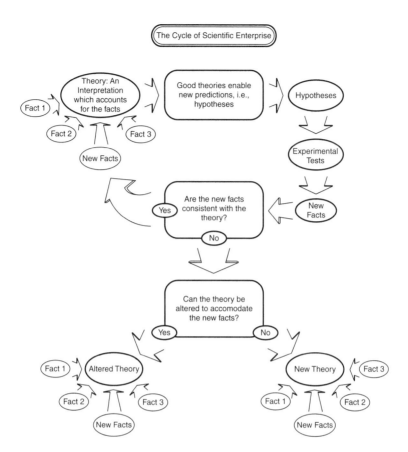

theories and the generation of new hypotheses, which the Scientific Method does not do.

The heart of the Cycle of Scientific Enterprise is the scientific theory. As defined above, a theory is not a single idea, nor is it just a guess or a hunch. Theories may begin as simple ideas, but for a theory ever to see the light of day in the scientific community it must be developed into a detailed explanatory framework that accounts for and explains what we know and gives us guidance on how to learn more.

Thus, there are two primary characteristics theories must possess. First, a good theory must account for all, or nearly all, of the scientific facts related to it. The theory must explain and make sense of the facts by describing how they fit together and what they mean. Ideally, a theory would account for all known facts. In practice, this is seldom the case, and we usually have some facts that are explained by the theory, some that we think will eventually be explained by the theory, and some that cannot be explained by any natural process that we know about yet. Second, a successful theory must enable new hypotheses to be formed from it. In other words, the theory must enable new predictions to be made and successfully tested.[7]

The Copernican Revolution of the sixteenth and seventeenth centuries provides us with an excellent example of how theories work. Prior to Copernicus' new theory of a heliocentric solar system (published publicly upon his death in 1543), the reigning theory on how the heavens worked was the Ptolemaic geocentric model. The Ptolemaic theory worked amazingly well. It accounted for the movements of the known heavenly bodies, it explained why planets appeared in different parts of the sky at different times, and it explained why the nightly progress of a planet against the steady background of stars could at times reverse direction (retrograde motion). Further, the mathematics of the Ptolemaic model were quite sophisticated and could be used to make predictions about planetary movements, such as dates for eclipses and planetary conjunctions (alignments).

Copernicus' new model would have gone nowhere if it did not do at least as good a job of explaining the facts and making predictions as the Ptolemaic model did. In fact, the Copernican theory did do as well as the Ptolemaic theory, accounting for retrograde motion and all the rest. And since the mathematics were significantly simpler (for the simple reason that the model correctly held that the sun was in the middle of the solar system rather than the earth) it quickly drew the interest of astronomers. Predictions made using the Ptolemaic model always contained some error— a predicted eclipse, for example, might be off by a few days or weeks. With the simpler mathematics of the Copernican model predictions still had some error, but they were comparable in accuracy to predictions made from the Ptolemaic model.

---

7 Keith Miller adds two additional characteristics good theories must possess: aesthetics (beauty, symmetry, simplicity); and fruitfulness, meaning that the theory must suggest new pathways for further exploration (Keith Miller ed, *Perspectives on an Evolving Creation*, p. 7). A friend of mine suggests yet another: Theories must exhibit coherence with other accepted theories.

Tycho Brahe used the Copernican model to predict the 1563 planetary conjunction between Saturn and Jupiter. This new prediction turned out to be close but still off by a few days, which told Brahe that even though the Copernican model appeared to be an improvement over the Ptolemaic model, it was still not completely correct. But in 1609 Johannes Kepler resolved this problem by discovering that the planetary orbits were elliptical, and not circular as Copernicus had thought. (Galileo stuck with the circular orbits to the bitter end.) With Kepler's improvement of elliptical orbits, the Copernican model of the heavens, which included an earth rotating on its axis and the moon orbiting the earth, became triumphant and stands to this day as our theoretical understanding of how the solar system works. Presently, our mental model of the solar system includes the heliocentric structure proposed by Copernicus with the elliptical orbits added by Kepler.

Let us now use this story to walk through the stages in the Cycle of Scientific Enterprise. Beginning with the facts that we know, we could make a long list: the sun rises and sets every day; the moon's shape appears differently every night in a cycle that lasts 28 days; Mars sometimes exhibits retrograde motion; the pattern of the stars appears to be fixed and unchanging, and so on. Any theory about the solar system (formerly known as "the heavens") must account for these facts (and many others) by relating them together into a single model. It is worth mentioning again that these days there are thousands of facts in any area of study, and it is probable that an emerging theory will not initially account for every one of them. But if the theory is any good, then over time the model, or a modified version of it, will explain an increasing number of facts with increasing confidence and clarity.

Once we have a theory that explains the facts, our Copernicus/Kepler model, we should be able to use it to form new hypotheses. If we cannot form new hypotheses, then scientific inquiry comes to a halt, because the way scientific knowledge progresses is through forming new hypotheses and testing them. Theories must be fruitful or they are sterile and useless. So let us form an hypotheses using this model:

Hypothesis: Sunrise in Austin, Texas on December 17, 2007 will occur at 7:22 a.m. CST.

This hypothesis contains all of the key elements in the definition above. It states a specific, measurable outcome that is expected (assuming the theory is sound), it is a positive proposition, and it can be tested by experiment. So now we are ready for the experiment, which is to wait for December 17, 2007 and to record what time sunrise occurs.

Let's say we do this and the sun rises as predicted. (Of course, it did.) What can we say about this? What we say is that the *hypothesis was confirmed*. Further, since the hypothesis was derived from our Copernicus/Kepler model, this theory has now been strengthened because of this success. If the theory is used repeatedly to make hypotheses that are successfully confirmed, then over time the theory becomes stronger and stronger because we begin to place increasing confidence in its explanatory power and in our ability to make correct predictions based on it.

It is very important also to note that this successful confirmation of the hypothesis *does not prove the theory*. In fact, we almost never speak of theories as having been proved, and if we do it is only after many decades of research and many thousands of experimental successes. As I mentioned in Chapter 1, in his 1922 Nobel Prize address Niels Bohr said that scientists "believe the existence of atoms to be proved beyond doubt." The *Encyclopedia Britannica* of the day said that "pure chemistry, even today, has no very conclusive arguments for the settlement of this controversy."[8] Today, of course, we would side with Bohr that we are very confident that matter is made of atoms. But we recognize that by his statement he only meant that we are as certain of the existence of atoms as we are of anything else in science. But as with all things scientific, the existence of atoms could, in principle, be called into question in the future.

What if the hypothesis had not been confirmed by the experiment? Does this disprove the theory? Absolutely not. A single experiment can never disprove a theory. There are many reasons why a particular hypothesis might not be confirmed:

- The experiment might have been flawed on account of an error in the experimental set up, an equipment malfunction,

---

8 James Gleick, *Genius*, p. 38.

or some unknown or unanticipated effect that influenced the outcome.
- The hypothesis might have been actually confirmed by the experiment, but the researchers misinterpreted the data.
- The hypothesis might not have been a properly formed prediction. In other words, the researchers might have misunderstood the theory and formed a bogus hypothesis as a result.
- The theory might be sound but incomplete. The hypothesis may address an area which the incomplete theory does not yet adequately encompass.

Because scientists and their equipment are fallible, mistakes happen. That is why the Scientific Method includes the "repeat the work" step. Experimental results, whether the hypothesis is confirmed or not, must always be confirmed by repetition. The more unusual the hypothesis is, the more a successful confirmation will need to be replicated before the scientific community begins to accept the result as legitimate. In these days of sophisticated, highly technical experiments, this typically means that the research is being conducted by a team of scientists who collaborate and check each other.

Research is also often conducted by more than one experimental team, perhaps in different countries, who compare their results, challenging or confirming one another. Many of us remember the announcement in March 1989 from two chemists at the University of Utah that they had discovered how to produce nuclear fusion in a jar at room temperature. Such an event would have rocked the scientific world, but as other labs tried to replicate the experiment scientists eventually concluded that nuclear fusion was not really happening after all.

Returning once again to our tour through the Cycle of Scientific Enterprise, we complete the cycle by considering how experimental results fit in with the theory that we started with. The results of the experiment, once replicated and vetted by the scientific community, become new facts which need to be accounted for by the theory. The hypothesis was derived from the theory in the first place, so if it was confirmed by the experiment, then the new facts (the results of the experiment) will be found to be consistent with the theory. The theory is now stronger, having given rise to a successful hypothesis and experiment, and thus now encompassing more facts in its explanatory framework.

However, if the hypothesis is not confirmed, and if after extensive scrutiny it appears that the hypothesis and the experiment were valid, then we would have to say that there are new facts that are not consistent with the theory. Again, this does not disprove the theory. When this happens scientists begin thinking that the theory may be incomplete and in need of expansion to encompass these new facts. Or perhaps certain aspects of the theory are in need of revision if this can be done while keeping the rest of the theory intact. By expansion or revision (or both) scientists will endeavor to develop a modified theory that does account for all of the old facts as well as the new results. If there is no way to modify the theory, then the door is open for competing theories to emerge that do account for all the facts. Over time, a new theory may gain credibility as it is used to develop successful hypotheses and the old theory may continue to fall on hard times as additional hypotheses drawn from it fail to be confirmed by experiment.

All of this can take a long time, decades or even centuries. Any major new theory, such as the Big Bang theory derived from Hubble's data on the expanding universe, Einstein's General Relativity, or Wegener's Plate Tectonics requires years of development and experimental success before the theory becomes accepted as the reigning model. And once a new theory has achieved such success, proving useful both in its explanatory power and its predictive power, it takes a very long time for the theory to be replaced by a better one, for the simple reason that a new theory would have to be demonstrably superior to the reigning model, both in explanatory power and predictive power, and this takes a long time to establish.

I used the Copernican Revolution as an example to illustrate the nature of theories and how they develop. A few more examples are in order to illustrate how hypotheses are drawn from theories. Perhaps the most stunning example of all time is the bizarre prediction Einstein made based on his 1915 Theory of General Relativity, that light from a star bends in space as it passes near another star. Einstein reasoned that if the light from a star were to pass near our sun on its way to us, the pathway of star light would bend and the star would appear to be shifted in the sky from its night time location when the sun was not near the path. Typically, we would not be able to investigate this prediction, because if the sun is near the path of the star light, that means it is daytime and we can't see the stars. But Einstein suggested that this effect should be visible to us during a solar eclipse. Sir Arthur Eddington jumped on this prediction and organized two teams of photographers to take photographs of the stars near the sun during the 1919 solar eclipse. Subsequent analysis showed the locations of stars whose light passed near the sun to be shifted by exactly the amount Einstein had predicted. This amazing experimental confirmation propelled Einstein to instant worldwide fame.[9]

Another very fine example of theory giving rise to hypothesis is Alfred Wegener's 1915 Theory of Continental Drift, also known as Plate Tectonics. This theory holds that the earth's crust consists of multiple moving "plates," nine major ones and at least as many minor ones, which drift around over time on the surface of the earth. An obvious hypothesis is that if the continents drifted in the past they are probably drifting now, which we could confirm if we were able to measure their velocities. At the time Wegener proposed his theory there were no technologies available to allow experimental verification that the continents were moving. But as laser and space-based technologies emerged in the 1970s several different techniques were developed and used to measure the continental velocities. All of these attempts succeeded, establishing that the drift velocities of the continents were in the neighborhood of one to two centimeters per year, and providing powerful support for Wegener's theory.

We have completed our tour of the Cycle of Scientific Enterprise, but there are a few additional comments that need to be made about theories.

First, as I hope I have adequately emphasized, it is incorrect to look down upon theories as inferior to facts. In the world of scientific knowledge everything is theoretically based. We never get away from the theoretical basis for our scientific knowledge, and to scoff at a scientific explanation as "just a theory" or "merely

---

9 This whole story is very engagingly told in Walter Isaacson's *Einstein: His Life and Universe*.

a theory" is to misunderstand how theories work in the cycle of scientific enterprise. Well-supported theories are the glory of science.

Second, laypeople tend to regard theories as having near-factual status after a massive number of confirmed hypotheses have accumulated over a long period of time. Our current belief that all matter is made of atoms is an example of a theory (the atomic theory of matter) that many people now regard as a fact. Other examples would be our understanding that mechanical energy and heat are simply different manifestations of the same thing, or that heating a gas makes the molecules in the gas move and vibrate more rapidly. These are all really theories, explanations developed to relate together all of the facts we know about atoms, energy, heat, and the motion of molecules. A hallmark of theories is that they are generalizations beyond experience, beyond our data. While it is correct to say that the matter we have explored appears always to consist of individual atoms, there is still a lot about matter we do not understand (such as, for example, how so-called "dark matter" fits in to our theories of cosmology). Thus the statement, "All matter is made of atoms" has to be regarded as a theory and not as a scientific fact.

Third, when discussing facts and theories with your students they may ask how "laws" fit in. Scientific "laws" are theories. In the early days of modern science we used to call theories "laws," but we don't any more. So what we call "laws" are just the old names that have stuck for theories we still accept.

Finally, sometimes a major shift, usually known now as a paradigm shift, occurs in the way scientists view the world. This is much bigger than having a new theory; it is an entire upheaval that changes the way we ask scientific questions. Paradigm shifts are rarer than the ongoing birth and death of theories, but many have occurred in the history of science. For example, when we ceased believing everything to be made of earth, wind, fire and water and understood the existence of many "elements," that was more than just a new theory about what things are made of; it was a paradigm shift.[10]

---

10 This is, of course, the subject of Thomas Kuhn's excellent study, *The Structure of Scientific Revolutions*, which gave rise to the phrase "paradigm shift."

Before concluding this Chapter we should consider why these scientific methods are as successful as they are. A short answer might be, "because they work." But from a Christian perspective, it is not just lucky for us that they work; it makes *sense* that they should work. God has chosen to govern the world through cause and effect relationships. He is an orderly Creator who has decreed that the creation will exhibit orderly, regular operation. In other words, natural processes operate in cause-and-effect relationships according to what we call the "laws of nature." The fact that they do so regularly and reliably means that scientific investigations are repeatable and over time we can amass a body of consistent scientific knowledge. Of course, since our knowledge is limited, our hypotheses are sometimes wrong, our theories don't hold up, and we have to keep searching for better explanations. But the fact that the world behaves so regularly give us the confidence we need to pursue scientific knowledge without the confusion that would result were the laws of nature to change every year or every day.

Another angle to this issue is prompted by the fact that so much of science consists in modeling the physical world mathematically. Why are we able to do this? Why *does* mathematics work? The layperson takes this for granted. But as many baffled scientists have pointed out over the decades, it is not at all obvious that mathematics, which consists of a set of relations we humans dreamed up in our minds, would have useful correspondence to the physical world out there, which does not reside in our minds (that is, if the world possesses objective reality, which almost all westerners accept).

As James Nickel has documented,[11] scientists and mathematicians have often remarked that the fact that mathematics can be used to model the physical world is something that borders on the miraculous, something we have no reason to expect, and something we should thank our lucky stars for. However, as Christians, we do have a straightforward explanation for why math works. The explanation begins with the truth that God created the world, and he causes it to operate with a regularity that lends itself to mathematical characterization. Next, God also created mankind, and endowed us with the mental capacity to imagine

---

11 James Nickel, *Mathematics: Is God Silent?*

mathematical propositions, and the curiosity that inclines us to try to understand nature and model it with the mathematical relations that we think up or discover. These are aspects of being created in his image. So there is a three-way dialog here. God made the world with certain properties, God made us with certain capabilities, and we live in the world and are curious about it. As far as I can tell, there is no other effective explanation for why math should be useful for modeling the physical world.

Application to the Classroom

The Cycle of Scientific Enterprise can be informally introduced in fourth or fifth grade, as teachers talk about theories and help students to begin to develop a proper understanding of them. Examples from different fields can be used to illustrate how our present scientific theories explain natural phenomena, and how hypotheses drawn from them can be experimentally tested.

When teaching students that hypotheses do not emerge from thin air but instead emerge from the scientist's theoretical perspective, the following example based on the dead battery scenario (discussed above) may be helpful. Tell students to imagine that they come from a planet that has no automobiles, and that knowing absolutely nothing about them they face the dead battery situation. Describe for students two or three completely separate "theories" of how automobiles work. One theory can be called the "animal theory." The animal theory holds that cars are living beings, animals, and that if the car won't start it is because it is unhappy. A second theory might be the "magic theory," which holds that cars run by magic. The third theory is the "mechanical theory," which holds that cars are machines made of carefully engineered systems of small parts. The group can then be challenged to come up with testable hypotheses from each of these theories. The animal theory might produce hypotheses such as "If we pet the car it will start" or "If we talk nicely to the car and leave sweets for it at night it will start." The magic theory will produce hypotheses such as "If we cast the right spell on the car it will start" or "If we wait for the moon to be in the right phase the car will start." The mechanical theory will produce hypotheses such as "If we replace the battery the car will start" or "If we give the car a tune up it will start."

When students enter high school they should be formally introduced to the entire Cycle of Scientific Enterprise, using a chart such as the figure above. They should at this time be required to learn clear definitions for the terms (fact, theory, hypothesis, and experiment) and they should be able explain how these terms relate to one another to make scientific progress possible.

It is unfortunately very common for students to have heard or used expressions such as "just a theory," or "merely a theory," as if the term theory referred to some easily dismissible bit of conjecture. Teachers should spend considerable effort to dispel this misunderstanding. Here are some examples of test items students can respond to:

1. Bill says to Sally: "I'm studying the theory of plate tectonics right now in school. It's really awesome because it explains how the continents were formed. According to the theory of plate tectonics, the processes that formed the continents in the first place are still going on!"

    Sally says to Bill: "Whoa! Get a grip! After all, plate tectonics is *just a theory*."

    Use what we have learned about to comment on the appropriateness of what Sally said to Bill.

2. If an experiment is performed and the hypothesis is not confirmed, does this mean the theory is disproved? Explain.

3. A scientist is heard to say, "As the first team to conduct this experiment, we were all excited about this research when our project began, but our experimental results were so disappointing that I am afraid our whole theory has been devastated." Using what you have learned about the Cycle of Scientific Enterprise, comment on this scientist's statement.

PART II

*Striving for integration and mastery,
with specific teaching tools
for the upper grades.*

Chapter 5

Verbal Matters

TO SET THE STAGE FOR THIS CHAPTER I NEED TO DEVOTE A COUPLE OF PARAGRAPHS to broad generalities about educational philosophy. I am generally annoyed when authors succumb to the temptation to write in broad generalities, so I would not do this if I didn't think it was necessary. Scores of books, ancient and contemporary, are devoted to the ideas that I am briefly going to try to summarize in the next few lines. But here we go.

The science department of a classical and Christian school is part of an institution that seeks to develop the whole mind of each of its students. This is one way to describe the essence of classical education, or liberal education, as opposed to a modern philosophy of education. One could say that modern educational philosophy boils down to viewing the purpose of education as to train people so they can be economically productive. But classical and Christian philosophy understands human beings not primarily as elements in an economy, but as beings with transcendent souls. The classical and biblical view of human beings is that our needs go far beyond college degrees and jobs. The purpose of education from a classical point of view is to lead students into wisdom and virtue through the pursuit of truth, goodness and beauty. I believe that this is also a biblical view of the purpose of education.

Modern educational philosophy places a priority on efficiency, and thus promotes the compartmentalization and specialization of knowledge. This results, for example, in science classes sticking to the business of scientific facts and laws to the exclusion of everything else. In modern schools the proper use of grammar is only required in Language Arts, the history of Europe is only discussed in history class, and the moral character of a researcher, if discussed at all, would be in the domain of a class on ethics (as is now the case in many colleges). But in a classically oriented philosophy of education human beings are regarded holistically. Classical philosophy affirms that knowledge in every field is fundamentally connected to knowledge in every other field and that education must develop, nurture and savor these connections. Thus, classically, scientists are first and foremost human beings, and being a good scientist means being a good human being who does science well. Good scientists must be of strong moral character, must respect other people, must be able to communicate clearly, must be able to apply mathematics when needed, must know their history, and, of course, must be good researchers who know their science.

In this Chapter I would like to focus on just one of the ways that studies in science must be integrated with studies in other fields: the linkage between science and language. Scientists do not merely solve problems. They must communicate with clarity to everyone else. Moreover, a student's understanding of a scientific principle is very clearly indicated by the coherence and lucidity with which he can articulate it. Students may claim (as they often do) that "I understand it, I just can't put it into words," but good teachers know better than to buy this argument. Show me a student who has understanding, and I'll show you a student who clearly demonstrates her understanding through language. It is essential, therefore, that in our science classes we hold students accountable for using language well.

Having set forth this basic claim, I wish now to outline a number of related pedagogical principles with some practical suggestions for putting each of them into practice.

- *First, assessments (quizzes, tests, papers, etc.) should always require students to express their knowledge in well-formed language.*

Student assessments in the ninth grade science courses where I teach are accomplished primarily through weekly quizzes (more on this later). On each quiz, 40-60% of the credit is attached to short answer questions targeted at students' knowledge of concepts, principles, applications of principles, historical events, and distinctions between key technical terms. Of course, exams in higher level courses such as physics and chemistry tend to be dominated by quantitative and symbolic material such as solving problems, balancing equations, and so on. But some space should always be devoted to having students express their understanding through language.

- *Second, assessments should use short answer questions that require students to write responses in a few properly worded complete sentences.*

On the verbal parts of quizzes and tests, use of matching, fill in the blank, and true-false items should be minimal or nil. Such items only test the lowest level of knowledge and do not require the student to exercise his or her mind in verbal interaction. In higher level courses teachers should avoid asking for simple definitions. Instead ask students to "distinguish between" or "compare" similar terms. Other useful verbs for verbal prompts are contrast, explain, describe, state, and characterize. Simple "Why?" questions are very useful: Why doesn't honey spoil? Why do ponds freeze from the top down? Why do metals bond with nonmetals? Students cannot give good answers to such questions without a solid grasp of the principles involved.

The ever-popular multiple choice test items, although easy to score, are problematic in two respects. First, as all the teaching literature confirms, they are hard to construct well and often are not very challenging. Second, even if they are constructed well and do present students with a challenging assessment of their conceptual understanding, multiple choice items do not require students to construct language for themselves, the need for which is the point of this Chapter. In fact, it is a general principle of pedagogy that learning requires the student to *construct*. This is why writing essays is such an effective way to learn. A student's mind must operate at a very high level of synthesizing information in order to construct a well-formed essay. Solving problems in mathematics and physics is the same way. Students always say that the problems look easy while the teacher is working them out, and while the teacher is working out the solutions, not much in the way of learning is taking place. But when students must begin from scratch to construct their own solutions, then real learning begins.[12] We will encounter this principle again when we discuss lab reports in later Chapters.

- *Third, in their written responses students should be held accountable for using correct grammar, spelling, and syntax at all times.*
  Additionally, they should be held accountable for clarity and logical coherence, which simply means that their answers are unambiguous, logically correct and make sense. A number of related points need to be made here:
  o Tell students to avoid pronouns like the plague. This helps eliminate vague references. Wait! Strike that last sentence and replace it with this one: Using nouns instead of pronouns helps eliminate vague references. See the difference? Additionally, cause-and-effect relationships and other logical connections between nouns need to be correctly stated for maximum clarity.
  o Prepositions matter, too. When stating or applying a physical principle, the difference between "with," "in," and "on" can make the difference between a student who has a solid grasp of the principle and one who does not.
  o Don't score a student's answer in terms of what he meant to say, even if you can discern what he meant (because of your knowledge of the subject matter and the discussions you had in class). Instead, score it by the literal sense of what he did say. Verbs, nouns, objects, and subjects must

---

12 I sometimes like to say, half-jokingly, that I would like to place a sign over my door that says, "No one ever learns anything in Mr. Mays' class." What I mean by this is that although in class I can direct students on to a road to learning and tell them what they need to learn, real learning happens when they try to work out solutions during their own study time, not by watching me do it on the board.

all be construed to render a statement that has plain, unambiguous, correct meaning.

o Don't allow what I call "shotgun answers." A shotgun answer occurs when a student packs his pen full of technical buzzwords and blows them randomly onto the page. Students do this when they don't really know or understand the answer to the question, but they do remember a few of the buzzwords from the Chapter that they heard in class, so they make up a worthless answer and pepper it with the buzzwords in the hope of getting at least a few points out of the question. The way to teach students to write meaningful answers and avoid shotgun answers is to give credit for the one and not for the other. Students will rise to the high expectations we set for them, as long as the expectations are reasonable and age appropriate.

o Don't give any credit for statements that do nothing toward answering the question, such as repeating the question, or for statements like "Galileo was a great man" or "Faraday's Law of Induction was an important discovery." Of course Galileo was great and the Law of Induction is important; otherwise we would not be studying them. When scoring students' papers simply strike through such statements and score the text that remains.

o Require students to answer the question that was asked. If a student's response leaves the question begging, then it isn't worth much no matter what it says. Reading a question correctly and responding to it is a basic requirement for learning how to use language well. When faced with a "why" question it is common for students to talk around the issue without addressing the "why" that was in the question. I sometimes give a certain amount of partial credit for this, depending on the nature of the question, the importance of the "why" part, and the amount time we devoted to the issue in class. As another example, if a student answers a question about Lavoisier by telling me a lot of correct information about Avogadro, I will usually award ten percent of the question's value. I am glad she knows the information about Avogadro, but the question was about Lavoisier and keeping them straight is important.

o Spelling matters, especially for technical terms that are part of the vocabulary of the lesson. I penalize every misspelled technical word, including common technical terms such as parallel, independent, ellipse, kinetic, and acceleration. For misspellings of nontechnical terms, I circle them while reading the student's answer, and then I judge from the number of circles whether the student is demonstrating an acceptable level of care in constructing his responses, or whether he is simply being sloppy (read: undisciplined). Anyone can make an occasional spelling error, especially on an exam, but a consistent lack of care indicated by frequent spelling errors needs to be penalized.

- *Fourth, grade verbal items by awarding points for excellence, rather than deducting points for errors.*

  We science and math teachers are accustomed to looking at a student's solution to a problem and deducting points for the errors. Partial credit is usually quite appropriate. If a student uses the correct equation and performs a calculation correctly except for one minor error, then deducting a point or two from a full score on the item is a customary and reasonable thing to do. However, your Humanities Department probably has a different approach for assessing student writing on themes and essays. In the Humanities Department at our school teachers score student papers by awarding points for merit rather than by deducting points for errors. Typically, a score of 70% means "satisfactory response." Scores above 70% are earned by students when they go beyond the merely satisfactory and move into the realm of excellence. Verbal questions in students' science classes should be scored the same way. An answer that contains no technical errors and is grammatically correct, but spare on detail should be judged satisfactory. Students who go further by giving full details and using high quality examples should be awarded the extra points for excellence.

- *Fifth, prepare students for success on verbal test questions by giving them practice at writing good responses.*

Lecturing on principles in class and instructing students to study their notes does not provide students with adequate preparation for success on verbal questions. Engaging in class question-and-answer sessions helps, but still falls short. For students to succeed, they need to practice writing the kinds of responses you expect.

The most successful and efficient technique I have discovered for addressing this is what I call the Daily Question. I inform my students about how this works at the beginning of the year, and then I use this frequently throughout the year. Once or twice a week I tell my students that it is time for the Daily Question. At this prompt they open their notebooks to a special section prepared for these questions. Then as I dictate the new question, which has emerged immediately from the lesson that day, they write it down. The questions I use are extremely similar (and sometimes identical) to questions students will see on their quizzes. Students are required to come to class the following day with a *written* response to the question (typically three to five sentences), an assignment that only takes a few minutes to complete. On the following day I select two or three students at random and ask them to read their responses aloud to the class. After hearing each student's response I critique it for the class. I begin by praising as many things as I can about the response, and then I comment on ways the response could have been improved. (This verbal interaction is a good opportunity to remind students to avoid pronouns in their written responses!) The whole exercise only takes a few minutes of class time, provides students with valuable practice, and requires no additional time grading.

To make the whole process work requires good management of a few more details. The first time we do this each year I encourage students to learn from my critique and not to be embarrassed by it. You might think this would be hard for 14-year-olds to do, but I have found differently. Students relish the opportunity to try their wings, to get praised for a well written response, and to hear specifically and directly how they can improve their answers. It's amazing, actually, how much they love it and how eagerly they write down my comments for improvement so that they will be better prepared in the future. Also, these responses are not turned in or graded. But since students do not want to be unable to respond in class if called upon, it is extremely rare for a student to come to class without having a response ready. If ever there is a student who comes to class more than once without being prepared, a simple email home to the parents instantly solves the problem. It is delightful to see how readily parents will lend support for student preparedness when the parents are paying thousands of dollars each year for the child's education.

- *Finally, standardize these expectations throughout your department.*
  Students must understand that good writing is expected in every science class in the school. The old adage that "a rising tide raises all boats" applies here. Performance in every class will improve when students understand that high verbal expectations apply not only in the class of one particularly picky teacher, but in all of the courses offered in the department.

Chapter 6

Quantitative Matters

No RESPONSIBLE SCIENCE EDUCATOR IS GOING TO ARGUE WITH THE PROPOSITION that mathematics is important for studying science. We all love quoting Galileo on this:

> *Philosophy is written in this grand book, the universe, which stands continually open to our gaze. But the book cannot be understood unless one first learns to comprehend the language and read the letters in which it is composed. It is written in the language of mathematics, and its characters are triangles, circles, and other geometric figures without which it is humanly impossible to understand a single word of it; without these, one wanders around in a dark labyrinth.*[13]

But despite the fact that we all agree that knowing mathematics is indispensable for learning science, we still face the ubiquitous problem that students struggle with mathematics, and they struggle with applying mathematics in science. It is not my intention here to dwell on the 50-year-old controversy in our country about students' skills in mathematics. The National Assessment

---

13 Galileo, *The Assayer*, 1623, p. 237-238, in *Discoveries and Opinions of Galileo*, Trans. Stillman Davies, New York: Doubleday, 1957.

of Educational Progress has been reporting significant gains in mathematics proficiencies for over a decade, while other studies routinely critique these assessments as flawed. This debate is never ending, and each new article one reads can leave one either optimistic or depressed.

But my own experiences teaching in seven different educational institutions (public, private, and collegiate) in Texas over the past 18 years have persuaded me of two crucial facts that are relevant here:

1. The actual mental capabilities of most students far exceed what we tend to see from them in their daily classroom performance; and
2. It is very, very easy to lead students through a year of study only to find at the end of the year that while they may have had a good time, they have not actually mastered any new skills and they will not take much with them into their new courses the following year.

As a professional educator I find these two facts to be endlessly challenging. In fact, I find the daily intellectual challenge of solving the problem posed by these two statements to be extremely compelling. I think about it all the time. The first statement informs me that regardless of how well students performed in previous generations and what people say about how they perform today by comparison, the fact is that the average student can do a lot more than what he typically does do. This is both fascinating and exciting! It's like owning a race car and discovering that it is capable of much higher performance than it presently exhibits. The owner of such a car will become devoted to discovering how to draw out this untapped potential. I am absolutely certain that such is the case with many of the students I get in my classroom year after year.

But lest I get intoxicated with the excitement of higher achievement, the second statement sobers me up by reminding me of just how challenging this will be. Many is the time that I have seen students arrive at final exam review week with only the weakest memory of topics we had studied in depth only a few months back!

Now to return to the topic at hand: When it comes to mathematics in science classes we need to discover the answers to two fundamental questions:

1. What is the overlap between a) what students *can* do, and b) what students *need* to be able to do?; and
2. Once we discover what these things are, how do we get students to master these skills so that they can use them in other studies in the future, gaining in proficiency and scientific capability year by year?

The problems of mastery and retention, which are addressed by the second question, are part of what I call the "Cram-Pass-Forget" cycle. Students cram for a test, pass it, and then three weeks later they remember very little of what they learned. (Every time I describe this to another teacher he or she knows exactly what I am talking about.) Come final exam week they try to cram it all in one last heroic time in a week-long blitz of studying. Then over the summer they forget most of it for good. Finally, after doing this annually for eight years, we have seniors in Biology 2 who believe tissue cells in plants and animals to be smaller than atoms (an actual instance). I will address the daunting challenge of mastery and retention, which I believe to be of utmost importance for effective education, in the next Chapter.

Returning to the first question, 13 of the 21 tools of science listed in the first Chapter are quantitative in nature. These are skills that I believe are both accessible to and essential for secondary students, but they are underemphasized or completely neglected in many secondary science programs. I will repeat each of them here and comment on them individually.

- *The student is knowledgeable of different units of measurement, including the USCS, SI and MKS[14] unit systems, and is proficient at performing unit conversions.*

  It is highly likely that if you teach in the high school grades you have been just as astonished as I have to see how many

---

14 These are the US Customary System (USCS), the *Système Internationale* (SI, typically known in America as the metric system), and the meter-kilogram-second (MKS) system.

students think that there are 1000 kilograms in one gram, or 60 seconds in one hour, or 100 cubic centimeters in one cubic meter. If you want them to be able to convert between U.S. Customary Units and S.I. units, you have probably also been amazed at how frequently students will say that there are 1.609 meters in one mile. When confronted with these errors students typically groan aloud at the silliness of their own mistakes, and say that they knew better and they can't believe they did that. The cubic centimeter to cubic meter conversion may require a couple of minutes of explanation the first time, but after that when students make that mistake they always say, "Oh yeah! I remember you told us that! I can't believe I did that!" Now, as I said before, we will address the issue of retention and mastery in the next Chapter. The point here is that it should be obvious that no student who makes unit conversion errors like the three examples listed above can be considered to have mastered his scientific studies. Moreover, performing unit conversions is one of the most frequent quantitative tasks in science, so if there is any task students must master it needs to be this one.

But schools do not approach unit conversions this way. I clearly remember in my own case that I had never even heard of unit conversions until I was a junior in high school taking chemistry, and the process of figuring out which factors to use and which way to write them was a complete mystery to me for months. Clearly, the school systems I went through as a youth failed me in this area.

Facility at performing unit conversions needs to be as familiar to our students as sending text messages on their mobile phones. (I don't do texting, but from what I have seen performing unit conversions may be the simpler task!) They need to begin learning conversion concepts and systems of units, and memorizing basic metric prefixes and standard unit equivalences in grammar school. Science courses in seventh grade and up need constantly to be building on these basics. In ninth grade students should be introduced to the MKS (meter-kilogram-second) system, a subset of the SI (metric) system, while maintaining facility with basic intersystem conversions such as meter-mile, meter-foot, and centimeter-inch.

Ninth grade, by which time most students have had or are in algebra, is also the time to require students to remember how to convert between degrees Fahrenheit, degrees Celsius, and Kelvins using a couple of memorized conversion equations and algebra to figure out the rest. In high school science courses virtually every quiz or test should include problems that require students to perform unit conversions. Moreover, these conversions should not be stand-alone exercises; they should be embedded in other computations, like they are in a real-life context. If you want a student to compute the kinetic energy of a baseball, give the mass in grams and the velocity in miles per hour and require the student to convert both to MKS units before using the kinetic energy formula.

Of course, simply indicating the correct units of measure that go with an answer is also very important. I always deduct one point from any answer that does not have the correct units of measure indicated with correct symbols. (I assign the same penalty in my math classes for application problems on quizzes and tests.) Every now and then I have a class of students who are happy to ignore the units if it is only going to cost one point per instance, in which case I tell the class that either they will become more conscientious about it or I will raise the penalty to two points or more. This usually gets the results I am seeking.

- *The student can describe the mathematical relationships embedded in the equations that express physical laws. Such relationships include direct and inverse proportion, independent and dependent variables, the behavior of different variables under different conditions, and linear and nonlinear functional relationships.*

If you have spent any time around working Ph.D. scientists in the graduate college at a university, you know that statements such as these are as common as paperclips:

"We were gratified to see a simple linear trend in our data, resulting in a directly proportional model with constant of proportionality equal to 3/2."

"Our research was based on the basic principle that the rate of heat flow varies as ΔT to the fourth."

"Well, of course the volume increases much faster than the surface area, since it varies as R cubed."

"The gravitational force is weak at long distances since it varies as the inverse square of the distance."

The fact is that research into the laws of nature often takes the form of a search for the mathematical relationships between two or more variables, and scientists speak of these relationships all the time. This is the language of variation and proportion. I read somewhere years ago that Paul Dirac once said, "I understand what an equation means if I have a way of figuring out the characteristics of its solution without actually solving it." Enabling students to think this way is challenging, but feasible and necessary if they are to gain insight into what it means to model the natural world mathematically. However, I have never seen a single science text that tries overtly to teach students to think in the quantitative terms of variation and proportion. This is one of those many instances in which the tools of science are just as important as any particular physical principle, but textbook authors seem to think that students will somehow pick up this language on their own. Some do, of course, but I maintain that we need to lead all of our students into being able to think in these terms.

Toward this end I developed an entire unit of study on this topic for my accelerated ninth graders. I teach them about a number of common types of variation (direct proportion, inverse proportion, square, inverse square, and so on), the terms we use (independent variable, dependent variable, constant of proportionality), the difference between the effects in equations of constants, variables and powers, and how to recognize and describe the way one variable in an equation will vary with respect to another one. This study consists of nine separate activities, each of which explores a different type of variation. For example, I use Newton's Law of Universal Gravitation, $F = G \frac{m_1 m_2}{r^2}$, as a case study for exploring inverse square variation. Students make a careful graph of F vs. r using the actual value for G (which requires them to consider how to handle the extreme powers of ten that occur in their tables of values and their graphs). Then they "normalize" the equation to $F \propto \frac{1}{r^2}$, plot this equation on a separate set of axes, and compare the two curves. After completing the activities most students are able to consider two variables in an arbitrary equation and describe how one varies with respect to the other using conventional phrases such as "the force of gravity varies as the inverse square of the distance between the centers of the objects."

- *The student knows the difference between accuracy and precision and how both of these relate to taking measurements.*

  Students should be introduced to this issue early in their high school science courses and it should be developed constantly year to year after that. By their freshman year honors or accelerated students should begin learning how to count up the significant digits in a measurement and how to apply the basic rules for how many significant digits should be quoted in the result of a calculation. Students in regular (non-honors) courses should learn about significant digits no later than chemistry. This is a difficult topic for most students to comprehend (which is why most courses prior to chemistry ignore it), but it is crucial. The difficulty students have is compounded by the fact that the terms accuracy and precision are constantly used interchangeably or incorrectly in common speech, and even in many texts! However, I have found a couple of techniques that are helpful when teaching this.

  First, motivate students by telling them that the precision issue is the answer to the question they always have about how many of the digits from the calculator display they should write down. Virtually all students are distressed or annoyed by the 10 digits in a calculator display and are eager to learn how to know how many of them they need to use. Second, since typical lists of the rules for determining which digits in a number are significant is somewhat unwieldy, students will find it helpful to know that if a number is put into scientific notation, then the digits in the stem (or mantissa) prior to the power of ten are the significant ones.

  This tip is not by itself a complete rule, but it really helps students to get a handle on the subject. So for example, I give students a number like 20,300 and ask them how many significant digits it has. The easiest way to get a handle on all those zeros is to write the value as 2.03×10^4, from which we can see that the value has three significant digits. This same helpful aid works for a number like 0.04055, which when written as 4.055×10^{-2} is seen to have 4 significant digits. Of course, with a number like 0.0075000, one has to go back to the rules to see that this value has 5 significant digits.

While we are on this topic I will pass on the most succinct complete rule for counting significant digits I have encountered:

The number of significant digits (or figures) in a number is found by counting all the digits from left to right beginning with the first nonzero digit on the left. When no decimal is present, trailing zeros are not considered significant.[15]

Beginning with chemistry, I deduct one point per instance when students' answers do not contain the correct number of significant digits, based on the information they were given and the conversion factors they had to use during the computation. For accelerated students I begin applying this rule in ninth grade.

Laboratory exercises which require students to take measurements from analog instruments (rather than merely reading a digital display) are a primary activity for driving this point home. It is in the context of taking measurements that students should learn that accuracy is about avoiding error, while precision is about reading the correct number of significant digits from an instrument.

- *The student can formulate a quantitative hypothesis.*

Nearly all of the laboratory activities in physics, and many of those in chemistry, require student teams to seek some quantifiable result. (Many lab activities in biology are descriptive in nature, and do not involve the formation of hypotheses.) In these experiments students should learn to form quantitative hypotheses that are structured

---

15 Charles McKeague and Mark Turner, *Trigonometry*, 6th ed.

the way hypotheses are in the real world of scientific inquiry. As an example, Einstein's hypothesis for the shift of the apparent location of a star during a solar eclipse (see Chapter 4) could be written this way:

According to the Theory of General Relativity, during a solar eclipse the apparent position of a star near the sun will shift away from the sun by 1.7 seconds of arc relative to its usual position.[16]

This hypothesis clearly states the theory from which the hypothesis is drawn, and makes a quantitative prediction that can be experimentally verified. Since predicting an actual value typically requires careful calculations which students may not have had the opportunity to perform in advance of the lab, I often encourage students to use hypotheses structured as in the following example:

Our team's hypothesis was that we would be able to use the principles of conservation of energy to accurately predict the speed of the toy car when it reached the bottom of the ramp.

Subsequent to the lab activity then, students perform the calculations, calculate the experimental error, and consider the question of how much error should be expected based on what kind of equipment was used.

- *The student can manipulate data, represent data in tables and graphs, and compute experimental error.*

These skills are very basic indeed, and in a well-designed lesson they are not difficult for students to learn. However, a very common barrier to students' acquiring these skills is the use of canned lab manuals, the type available from nearly every publisher and used ubiquitously in secondary science programs. These lab manuals have a separate section for each lab activity, with perforated pages so students can tear them out and turn them in for their reports. In addition to procedural instructions and "thought questions," the pages include blank tables and graphs for students to use.

---

16   This figure was obtained from Walter Isaacson, *Einstein*, p. 256.

As I mentioned in the previous Chapter, learning occurs when we construct or synthesize content for ourselves, not when we fill in the blanks in a form developed by someone else. For this reason, I strongly maintain that we should not use these published lab manuals at all. This means, of course, that we have to develop our own materials to give students the instructions they need and to show them how to write a good lab report. We will discuss laboratory work at more length in Chapter 9, but here we will consider a couple of examples pertaining to the skills involved.

As a first example, let's say a student is collecting data for a lab on Newton's Second Law of Motion. She rigs up some system that will apply a constant force to small car and she measures the time it takes the car to traverse a certain distance. From these data she calculates the car's acceleration. To explore the linear relationship between the acceleration and the force, she performs the experiment using three different amounts of force, and to make sure her data are consistent she performs three separate trials at each force value. Now it is time to display the data in the report. A published lab manual will have in it a table similar to this one:

mass of car: _____ grams

distance: _____ centimeters

Time Data (enter times in seconds)

	Force		
Trial	1.0 N	2.0 N	4.0 N
1			
2			
3			
Average			

Presented with a blank table like this, the student doesn't have to think at all about how to present the data. The thinking has all been done for her. How to arrange and label the nine time values and where to put the average times are already decided. Moreover, the student doesn't have to remember to indicate the units of measure, this is already done, and she also doesn't have to remember to report the mass of the car and the length of the timing zone, because there are blanks for these values, too. In other words, the student simply doesn't have to think at all when filling out a form like this. On her own, the student would have to think through the different ways to lay out this table. She would have to remember to display the units of measure using the correct symbols. And she would have to remember to report the miscellaneous data values such as the mass and the distance, which students often forget to do when they are fussing over the time data, which seems to them to be the most important. The pedagogical value of requiring students to set up their own tables from scratch is huge. It is the difference between thought and no thought.

The same principles apply to setting up graphs. Canned lab manuals typically have a blank graph all set up and ready for students to plot points, such as the one below.

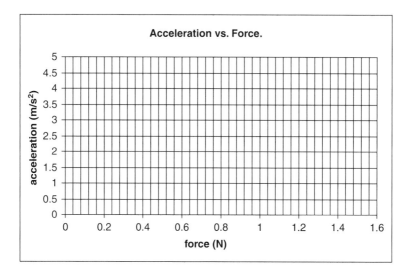

Clearly, as in the case of the data table, the student doesn't have to think much at all to plot points on this graph. But the real value in the exercise is in the very construction of the graph, where students have to decide for themselves what scales to use, which variable goes on which axis, what the units will be, and what to title the graph. Once again, students miss some of the most important lessons the lab has to offer when these formatting considerations are done by textbook publishers instead of the students themselves.

- *The student can identify reasonable and significant sources of experimental error.*

As we discussed in Chapter 4, experimental science progresses by making predictions (hypotheses) from theories, and then testing the predictions to see how close the results match the predictions. A critical part of this process is quantifying the difference between the predicted value and the experimental value. In secondary science courses this quantification is usually called the experimental error.[17]

As part of their discussions in lab reports students should be required to calculate the experimental error and identify potential sources for that error. Accomplishing this much with the average student is not difficult. What is much more difficult is to lead students to begin thinking quantitatively during their speculations on causes of error. Without effective instruction on this students grasp at straws just to fill up the discussion sections in their reports, often suggesting sources of error that could not possibly have had any major bearing on the outcome of the experiment.

When critiquing the inappropriate remarks students make in their discussions the easiest one to deal with, which is also one of the most common, is a student's claim that the

---

17 Although the phrase experimental error is common in secondary science texts, it is somewhat of a misnomer, implying error in the experiment, even if known factors contributing to the experimental outcome (such as friction) were intentionally left out of the prediction. This phrase is not used this way in scientific research, for the simple reason that serious experiments are designed to take all known factors into account so that the experimental result will be as close as possible to the prediction. I will continue to use the phrase experimental error, but we need to make sure our students understand what is implied in the phrase and where we have departed from theory when we make simplified predictions.

Another aspect of error in measurements is the experimental uncertainty, which is present in every measurement, and relates statistically to the distribution in the measurements (σ). Upper level students in advanced classes should learn to apply basic statistics to their data to establish the uncertainty in their outcomes. This is discussed briefly below, and again in Chapter 9.

experimental error might have been caused by errors in his calculation of the predicted value. I tell students that there should not be such errors in a finished report and that they can't blame their experimental error on their own calculations. Students work in teams, and the entire team should work together. If four team members independently calculate the same predicted value, it is unlikely that they all made an identical error in the process. If any team member computes a different value, then all team members should review the calculations together until everyone agrees on the correct way to perform the calculation and on the correct result. If after this group effort students are still nervous about the correctness of their predictions, then it is easy enough to confer with another group to see if the other group's calculations are consistent with their own.

A much more difficult mental bridge to cross is enabling students to understand that the source of error they are suggesting has to have the mathematical capability of creating the kind of error they are seeing. For example, let's say students are performing a conservation of energy lab in which they use the mass and elevation of a toy car to predict what the velocity of the car will be when it reaches the bottom of a hill. Let's say that the apparatus used results in an experimental error of 6%. (Actually, with good Hot Wheels® cars and digital photo gates to clock the car at the bottom of the hill errors below 1% can be achieved.) When searching for explanations for this experimental error it is common for students to suggest that an error in their measurement of the height could have been at fault. The apparatus I use for this experiment involves a height difference of around 20 cm, and students measure this with a meter stick. I would say that the worst error a student could possibly make when measuring this height would be half a centimeter, with most student measurements agreeing with one another to within a couple of millimeters. Assuming the worst, a 5 mm error in a 20 cm measurement gives an error in the height of $\frac{|20-20.5|}{20} \times 100\% = 2.5\%$. Now, since the kinetic energy of the car at the bottom of the hill depends linearly on the height from which the car started, a good starting point for students to consider is if a small error (2 mm or so) in their height measurement can cause an experimental error of 6%. It can't, and they should be led to see this.

But older students should be pressed to think about this even further. The end of the experiment is to predict the speed of the car at the bottom of the hill, and predicting the car's speed from its energy requires taking the square root of the energy. If due to a 2.5% measurement error in the height our energy figure is at 102.5% of the value it should be, then taking the square root of this will produce a speed prediction that is 101.2% of a prediction based on the correct height. In sum, a height error of 2.5% produces an error in the speed prediction of only 1.2%, and thus this error source is not very useful when trying to account for four or five times this much error. (If the measurement were 2.5% below the correct value the predicted speed would be in error by 1.5%.) In reality, it is friction between the car and the track that accounts for almost all of the experimental error. However, students should not automatically attribute the error to friction either. If the experimentally determined speed is above the predicted value, then friction certainly cannot explain the error. But if the experimental speed value is below the prediction then attributing the error to friction would be reasonable.

I do not believe that students can really begin completely to master this more advanced analysis until their upper grades in high school. But students do not learn to think this way in a short amount of time. It takes repeated experience in different experimental contexts in different groups over an extended period of time for these ideas to sink in. For this reason, it is important that students are introduced to this kind of analysis no later than ninth grade so that these concepts have time to grow during the course of their high school studies as students come back to them several times each year.

- *The student can estimate the uncertainty in a measurement.*

  Upper level students should encounter the reality that every measurement contains uncertainty. With good equipment and careful techniques the distribution in a data set should be apparent. Students who have had a course in pre-calculus should have learned how to calculate the standard deviation of a data set (s or σ) and can use this value as an estimate of the uncertainty. The graphs in their lab reports can include error bars to reflect the standard deviation for each set of trials.

- *The student can set up a scientific graph with appropriate labels and scales, and can use it to compare theoretical and experimental values and trends.*

Graphing is one of the most useful and universal tools in science and math, and the skill of creating proper scientific graphs is crucial. Science activities and lab reports should engage students in graphing results of calculations or experiments as often as possible. Moreover, every time a student constructs a graph, even on the most routine homework assignment or the most perfunctory quiz, the expectations for the student's work need to include each of the following:

o The two variable names on the graph's axes are labeled.
o The units for each of the two variables are given, using one of the two standard formats, as in these examples:

acceleration (m/s^2)

or

acceleration, m/s^2

o Numerical scales on both axes are shown (and are, typically, linear).
o The graph has a title.
o The graph is well proportioned. That is, the scales were chosen properly to make the graph easy to read.
o The graph is drawn with neatness and care, and is reasonably accurate.

It is also important that students learn how to develop a curve representing predicted values and display it along with a curve displaying the experimental values. Students in accelerated science classes should learn to do this in the fall semester of their freshman year, and should simultaneously learn how to use Microsoft Excel, or a similar tool, to create a graph with both curves and insert this graph into their typed lab reports.

- *The student can identify outlying data and suggest alternatives for dealing with outlying data.*

  This is a more advanced skill. At the high school level most students will not have had the background in statistics necessary to apply statistical methods to treating outliers in a data set. If upper level students are in a statistics course, then it would be desirable to leverage that instruction and apply it to experiments in their science classes. However, even without the training in statistics all students should encounter more basic methods such as repeating trials until several measurements fall within a certain range of each other and throwing out the ones that do not, as long as it is clear from the procedure that the "good" measurements are definitely clustering within a reasonable distribution. Another basic method is to simply throw out a data point when the experimenters are certain that the trial was contaminated or rendered invalid for some reason.

- *The student can confidently identify reasonable solution strategies for problems when a method is not readily apparent.*

  This is also a more advanced skill, but it is essential to success in chemistry, and it is particularly essential for success in physics. A standard problem of this type that students encounter in physics is the following:

  > *From what height would a lead bullet, mass 8.5 g, have to be dropped in order for it to melt completely upon impact with the sidewalk below? (Assume no air friction, that the bullet has a temperature of 20° C at the moment of impact, and that all of its kinetic energy is converted into heat energy in the bullet.)*

  Many students encountering this problem for the first time have no idea how to proceed. But this is frequently the way it is in science, and learning to imagine how what one knows can be used to figure out what one does not know is a very valuable tool to have in one's cognitive toolbox. Teachers need to do two things here. First, we need to make sure that students encounter plenty of such problems (even though life becomes more difficult for everyone when they do). Second, when students ask us how to solve the problem we need to refrain from telling them. The value of the problem is lost if the student is told how to solve it. Instead, we need gently to suggest that the information given in the problem contains clues to its solution. Further, we might suggest that the student consider that the various topics we have studied (energy, thermodynamics, calorimetry) must be related to one another in order to solve this problem. Finally, we have to insist that the student spend a significant amount of time studying the problem and trying different things before giving up. The value of the problem lies particularly in this mental effort. If the student is told that he should calculate the amount of energy it takes to melt the bullet and then use this as the kinetic energy of the bullet, and then use conservation of energy to equate gravitational potential energy to kinetic energy, then the real value of the problem as an exercise in critical thinking is lost, because the problem is reduced to plugging numbers into formulas after that.

- *The student can proficiently apply mathematics to the solution of problems.*

    Perhaps the need for this important skill goes without saying. But the fact remains that it is common for students to be able to solve equations in their math classes and then get stuck when similar equations appear in their science classes. I suspect that this phenomenon is an artifact of the modern educational tendency toward compartmentalization, which I referred to at the beginning of the previous Chapter. Students think that math only occurs inside the four walls of the math classroom and they are stymied when they discover that they need to be ready to use their math skills elsewhere. Addressing this problem is straightforward, but sustained efforts over a long period of time are necessary before the compartmentalization mentality is broken down throughout the school. Math classes need to use many, many application problems drawn from as many different fields as possible. Moreover, these application problems need to be as realistic as possible. On the science end, science classes need to pack in as many quantitative problems as they possibly can. Science teachers need overtly to call students' attention to the fact that the linear equations they are encountering in kinematics, or the power functions they have to solve in chemistry, or the exponential growth problems they face in biology are the same functional forms they have already studied in mathematics.

Even more importantly perhaps, science teachers need to insist that students demonstrate mathematical competency as one of the criteria for good grades. This is no small issue. Across our country science and math courses are often diluted or dumbed-down to make them accessible. The teaching literature is full of material about presenting students with material that relates to their life experiences, but articles on holding the line on mathematical competency are comparatively rare. (The previous two sentences would be regarded as scandalous in many contemporary educators' circles, but I stand by them.) Moreover, if a teacher holds the line on mathematical competency in a science class it is highly likely that he or she will face strong criticism from parents for making the class too hard. Parents will complain that their child has never made a grade lower than A in his life until he arrived in this class, and that it is unreasonable to expect students to remember the things they are taught. (No, I am not kidding.) Again, we will hit the issue of retention in the next Chapter. But here I am affirming that being competent in science includes being able to use mathematics in science. Instead of trying to avoid this reality, we science teachers need to glory in it and joyfully lead our students into the intellectually satisfying world of being able to use math to solve problems in science.

- *The student can use mental math to get quick approximate answers.*

Every educated adult I know over the age of forty is amazed that young people can't seem to figure out a 15% tip at a restaurant. Every math and science teacher I know, and nearly every parent I know, believes that students should be able to estimate numerical results. But leading students into competence in this area is like trying to bathe cats. It seems like it should be so *obvious* that $\frac{\pi}{3}$ is just a little more than one, that e^2 is less than 9, that $\frac{20 \times 10^6}{9.87 \times 10^5}$ is about 20, and that 10% of the speed of light is *not* 2.7×10^8 meters per second. The fact is, these estimations are not obvious to most students. Maybe 100 years ago every 8-year old was better at mental math than students are now because they had to figure up their earnings on the newspapers they sold (like my grandfather did). Maybe our students are more challenged now because they routinely depend on calculators to do their thinking for them, just like students of literature are more challenged now because people don't read books like they used to. There are hundreds of books on these topics, and for every conservative who says students are worse nowadays than they used to be there is a progressive who claims the opposite. I'm not going to get into that controversy.

I do know this: Estimating numerical answers is not that hard to learn. Like everything else, it just takes practice. Also like everything else, it cannot be "learned" in a two-week unit on estimating and then not touched again in the curriculum. This is worse than not covering it at all because it was a waste of everyone's time. The key to learning and mastering anything is to do it frequently. Students should begin learning to estimate in Grammar School, and should be required to practice estimation frequently ever after. This requires a significant amount of cooperation between the math and science departments in the school, and a considerable amount of creativity and intentionality on the part of the teachers.

Science (and math) teachers in grades 7-9 should particularly emphasize exercises such as those listed above. And while they are doing it, this is the perfect time to practice reductions of units of measure. Construct problems such as $\frac{20 \times 10^6 \frac{m}{s} \cdot 3kg}{9.87 \times 10^5 \frac{s}{m}}$, and while the students are reckoning the numerical value to be around 60, have them also reduce the units to $\frac{kg \cdot m^2}{s^2}$. Older students (high school) should be required to go one step further and identify these units as equivalent to joules (J). It is painfully common for students to divide 50 m/s by 2 s and figure that the seconds cancel, leaving meters as the units for the answer (instead of the correct m/s^2). In more advanced studies in science, such as physics, manipulating the units of measure can be a powerful problem solving aid, if students are proficient at manipulating units of measure to start with.

- *The student recognizes when calculated answers or experimental data are unreasonable or erroneous.*

  This skill follows from and is perhaps the most immediate application for the previous skill. A quick mental calculation can often be used to determine if a calculation looks reasonable. But even without using mental math students should be watchful for unreasonable results that indicate mistakes they have made. Any student might make a minor error when computing, say, the velocity of a baseball. But if the student's calculator error results in an answer of 2.3×10^{14} meters per second, and she already knows that the speed of light is 3.00×10^8 m/s, then the penalty for this error needs to be made more severe.

Chapter 7

Science as a Cumulative Discipline

IN THE LAST CHAPTER I REFERRED TO A PHENOMENON EVERY EXPERIENCED TEACHER knows about, what I call the Cram-Pass-Forget cycle: Students cram for a test, pass the test, and then proceed to forget most of what they "learned." I place "learned" in inverted commas because it is my contention that if one can only remember something for three weeks then one can hardly be said to have learned it at all. We all know that this situation is ubiquitous in American schools, but what are we doing about it? The educational literature these days is full of information about the quality of students' experiences, making the material relevant by relating it to the students' personal lives, and making the material practical by relating it to careers. But to my mind, a situation in which students do not remember much of what they have been taught is simply unacceptable. I will immediately qualify this assertion by acknowledging that I don't mean that every student should remember—for the rest of his life—100% of everything he learned in my class. That would be a preposterous standard. But common sense tells us that if a good student really *learned* a principle or concept, and was given repeated opportunities to rehearse the knowledge, then he should remember it well for an extended period of time.

In this respect studying science is more like studying mathematics or a foreign language, and less like studying literature or history. It is a cumulative discipline, in which new topics are based on and follow from principles learned in previous studies. To an extent, one can study a literary genre or a particular period of history without remembering many specifics about previously studied genres or periods. But one cannot study molecular biology without remembering chemistry. And one cannot excel in chemistry without remembering the basic principles (such as atomic structure and electrical attraction) learned in a physical science class. Moreover, the tools of scientific inquiry we have discussed in previous Chapters are tools required for every science course. They are like basic skills in mathematics, and attempts at further study will be ineffectual if these basic tools from previous studies are not mastered and retained.

So what should it mean for a student to make an A in your science class? I argue that it should mean that if one walks up to that student without warning in July and asks her a question that relates directly to the basic learning objectives that were addressed in the class during the previous year, there is a good chance she will still be able to give a decent answer to the question. That is what the A should mean—that actual learning, A-grade learning, really did occur in this student's case. The student learned the material and is ready to take it with her into her studies the following year.

If an A does not mean this, then what does it mean? It means that the student did a good job at completing all of the assignments and passing tests. But if the completion of those assignments does not result in the student *knowing* that material, why is the student doing them? To have a good experience? I can think of many ways to set up a class in which students would have good experiences, maybe even very enriching experiences, but would not learn very much of the discipline under study. I hope all my students have good experiences in my classes, but if this is all they take away from my classes then my pedagogy is a failure. Successful pedagogy results in students knowing valuable things they did not know before, and taking this knowledge with them to their future endeavors (further school science classes or otherwise). And when I say *knowing*, I mean they have comprehension of material and the ability to apply it that they will retain for a significant period of time.

I first began to focus on this issue many years ago when we arrived at final exam review week. In my ninth grade science class I passed out a list of topics that would be covered on the exam. One student's hand immediately went up. When called upon the student asked, "I see that kinetic energy calculations will be on the test. Did we cover that?" Many other students began nodding their support for this question, indicating that a significant portion of the class could not remember even covering this topic. The fact was that we had done what virtually every physical science class does. We had studied energy calculations for three weeks or so and had a test on it earlier in the year and had moved on to other material. The result was that many students hadn't learned a thing, and the ones that did learn something had only learned very little. So what does one do in this situation? One proceeds to the exam review periods during which the students try to learn in three days everything they should have learned in the previous semester (or year in this example, since the exam was supposed to be cumulative over the entire year). This being nearly impossible for most students to do, performance on the final exam in such a situation is typically poor unless the rigor of the exam is reduced to enable the students to pass.

Teachers should (and many do) deplore this entire situation. It represents a colossal breakdown in the quality of our schools and has led directly to the now common phenomenon of remedial instruction for entering college students. Accordingly, I have made it one of the central elements of my teaching to figure out how to structure my courses so that my students actually *learn* things. Of course, it is one thing to increase rigor so that students are compelled to remember material under the threat of a failing grade if they don't. It is another thing to enable them to manage and succeed under these higher expectations. Thus, we have two separate elements of pedagogy to discuss, pedagogical techniques that promote retention, and pedagogical techniques to enable students to manage these expectations and succeed.

Promoting Retention

The basic pedagogical tool kit for promoting retention must include these elements, which we will address one at a time:

1. Major assessments are cumulative.
2. Homework counts for very little credit toward the grade.
3. The frequency of major assessments is adapted to the grade the students are in.
4. Teachers in all courses set expectations for students to remember key information from previous courses.

Major assessments are cumulative.
This first principle is the *sine qua non* of this entire Chapter. The first thing to do when trying to break the Cram-Pass-Forget cycle is to make it impossible for students to succeed on tests this way. If your exams on Chapters 4, 6 and 8 all include review questions from Chapter 2, and if the students know that this is the case, then the students know that it won't do any good to cram for the Chapter 2 test and forget everything. Instead, they know they will have to remember it for every exam after that. If they did cram for Chapter 2, they will have to cram again for every test, but the amount that has to be crammed in increases by one Chapter every time. Students will eventually realize that they cannot deal with the course this way.

I hasten to add at this point that the students' knowing how this works is a necessary but not sufficient condition for success. They can't succeed in such an environment unless they know what is expected of them. But most students have no idea how to study properly for long-term retention. They know how to Cram, Pass and Forget. They can hold their attention on a topic for three weeks, and they can remember the equations, definitions, principles, and problem solving techniques for three weeks. After that, unless the structure of the course requires otherwise, they will forget at least 75% of what they were previously taught. Students do not automatically know how to study for retention. We have to teach them how. We will address this in the next section on "Enabling Students to Succeed in a Mastery Environment."

If you use exams for the major assessments, strive to make 20-25% of the material on the exam targeted at older material. If you use weekly quizzes (see below), roughly 25-50% of the quiz should be review material.

There are several ways to make tests cumulative. Use all of these techniques, separately or in combination:

- Put problems and questions on the test that come straight from the Objectives Lists for previous Chapters. (Objectives Lists are discussed in the next section.) Don't make these tricky or obtuse, and don't try to compel students on a test to think in ways they have never thought before. Do use straightforward problems right out of the material from previous Chapters.

- Ask questions that are new applications of previously learned material. For example, let's say your students study the water cycle in October, and in March they are studying various habitats and ecosystems such as ponds. You could ask them to describe how the water cycle affects the pond ecosystem under study. If new applications like this are rare, students will inevitably ask, "What is the water cycle?" But if such new applications are a regular feature of your teaching, both on tests and in everyday classroom instruction, students will know they are supposed to possess the knowledge necessary for fielding such questions and they will scramble to search their memories more thoroughly or they will look up the information on their own if they find they have forgotten it.

- Embed older skills into studies of newer ones. If your students have already learned unit conversions, scientific notation, and significant figures, then nearly 100% of the computations on your quizzes and tests should require students to use these skills. Textbooks are notoriously unhelpful on this point. Problems in texts tend to be stripped of the need to use any other skills except the specific ones under study. Thus, you will probably have to develop many of your own problems sets (or at least supplement those in the text).

In some upper level courses (primarily juniors and seniors) in which Chapter exams count for the major part of the grade (around 85%, see next section), quizzes covering only material currently under study can be used between major exams to assist students in keeping current and in finding out how prepared they are for the exam. These quizzes can be averaged together to count the equivalent of one exam grade in the semester average. But the Chapter exams themselves should all include older material, either explicitly or embedded in new problems and questions. For lower grades that use the weekly quiz (see below), every quiz should be cumulative from day one.

Homework counts for very little credit toward the grade.
I stated earlier that a student's grade in a class should indicate how much that student has learned. When it comes to assigning the grades, there is a critical distinction between those activities in which students must engage in order to learn, and legitimate assessments that measure how much learning has occurred. We all know that completing homework assignments is necessary if students are to learn. But it should be obvious that merely completing an assignment does not mean that a student has learned the material. There are many ways to get an assignment completed, including working together with peers, getting assistance from parents or tutors, studying a solutions key, looking up answers in a book, "reverse engineering" a computation so that the calculation arrives at the known correct answer, copying someone else's work, and so on. Homework is necessary for learning. But getting it done is not evidence that the student has learned anything.[18]

The most legitimate instrument for assessing how much students have learned is the in-class test, whether short quizzes or longer Chapter exams. For this reason, the vast majority of the student's grade (80-85%) needs to be based on cumulative quizzes and tests. Very little credit should be awarded for homework assignments (5-10% max for younger students; zero for older

---

18 The major exception to this rule is in writing original papers. An essay writing assignment compels students to engage material so closely that they must comprehend it well in order to write a good paper. But while writing an occasional essay in a science class may be very valuable, the objective nature of the course content dictates that essays cannot comprise a very large percentage of the students' grades.

students). The balance of the grade is based on the students' laboratory work and lab reports.

The question that occurs to many people after hearing the ideas expressed in the last two paragraphs is, "But if I don't give my students any grade credit for their homework, they won't do it." This argument simply does not hold up. Private schools have an extremely powerful tool for motivating students to do their work: communications with the parents. When parents are paying thousands of dollars each year to provide their child with a quality private education, the last thing they want to get is an email from the child's teacher informing them that their child did not turn in his assignment that day, or that it was incomplete or inadequate. So, literally, all one has to do is this: Tell the students that assignments must be completed so that they learn, but that it counts for little to nothing in the grade. The purpose for doing it is not to get a grade, it is to learn. Then get on with your class. When a student fails to submit an assignment on time send an email to the parents the same day informing them of it, and politely reminding them that their child cannot expect to succeed in your class if he does not complete his assignments. The problem has instantly become the student's problem with his parents, instead of your problem. This method works virtually every time. And if the student fails again in this area, just send another email.

Recognizing that maturity is a factor in the way students approach their work, I admit the benefit of granting a small amount of grade credit (5% or so) for assignments completed by students in grades 7-9. For students in grades K-6 the situation is a bit different. In these lower grades exams do not play the prominent role they do in higher grades. Instead, projects, participation and assignments are much more important, because at this level the student learns much more by participating, and much less by sitting at his desk with the books at night.

The frequency of major assessments is adapted to the grade the students are in.
After cumulative testing, the single most effective tool I have ever discovered for promoting retention among my ninth grade science students is the elimination of Chapter tests, replacing them

with a weekly 30-minute quiz. This one feature has revolutionized the amount of material my students retain, the facility with which they can handle computations even months after they learned them, and the depth of what they know. It has eliminated the need to review for final exams, since every single week all year long the quiz is a miniature final exam, comprehensive to the beginning of the year, and occurring week after week after week. I would sooner do without textbooks and classroom than give up the weekly quiz regimen. It works like a charm.

The weekly quiz occurs the same day each week at the beginning of class. I never have to announce it because it happens at the same time every week throughout the year. The students learn the drill the first week of school, and after that every week on quiz day when the bell rings they are ready and waiting, pencil and calculator at the ready. I pass out the quizzes and write on the board the time it will be 30 minutes later when the quiz ends. Each quiz covers all material studied in the course up through the preceding week. Since quizzes are cumulative, there is no correlation between material on quizzes and the dates new Chapters of study are initiated or completed. Whatever we have covered in the course by the end of a given week is fair game for the next week's quiz.

The front page of the quiz consists usually of four computational problems worth 15 points each. The back of the quiz consists of three to five questions that must be answered with a few complete sentences, or sometimes with a list or diagram. These mostly verbal items account for the other 40% of the quiz grade. The average of these weekly quizzes comprises 85% of a student's term grade in my class (prior to the contribution of the semester exam). Since each quiz is cumulative, each quiz becomes a study tool for future quizzes, and students take great interest in learning exactly why they lost points so they can improve their answers for the next time. Typically, the amount of content on a quiz that is new (appearing on a quiz for the first time) is in the neighborhood of 50-75%. Thus, roughly one quarter to half of the material on each quiz comes from topics we have studied in the past.

I began this Chapter with an anecdote about arriving at final exam review week only to discover that students couldn't even remember studying some of the important topics from the course. Since I have been using the weekly cumulative quiz not only do students remember almost everything, but the final exam review week itself is now superfluous. They have been reviewing everything every week for the weekly quizzes; there is nothing that needs to be reviewed for the final exam. Instead, we clarify any remaining questions and spend time drilling computations and rehearsing answers to verbal questions. Students can (and do) prepare for the final exam just like another weekly quiz (one that happens to be two hours long). There is nothing else like the cumulative weekly quiz for this age group (high school freshmen). It is a regimen that is perfectly suited to the way students at this age study and learn. It works so well that in my view the weekly quiz regimen is preferred for all science classes in grades 7-10.

As I mentioned before, students in lower grades (Grammar School) are not subjected to testing the way students are in grades 7-12. It might be a good idea to use a weekly quiz in Grammar School grades, but to make the quiz shorter, say 15 minutes long. For older students (grades eleven and twelve) the material they are studying is more sophisticated, the problems take longer to solve, and the questions more time to answer. In these grades the major test or Chapter test every three weeks or so is preferable, particularly since this is the way college courses are structured and students need to gear up for it. But in the critical adolescent years from grades 7-10 the weekly quiz is the way to go.

Teachers in all courses set expectations for students to remember key information from previous courses.
It is common for students to begin a new science course expecting that everything they need to know will be taught to them by the new course teacher. But if students are to graduate from high school with a solid background in science that they take with them to their future endeavors, teachers must expect their students to possess some knowledge as prerequisite to the new course under study. If only a single teacher adopts this attitude, then the teacher may be viewed as unreasonable or too demanding. But if the entire science department adopts this approach, then the students will view these expectations as normal and reasonable and they will rise to meet them.

In this respect, the traditional notion of the so-called "spiral curriculum" has done us a lot of harm. Instead of expecting students to learn things and take this knowledge with them into further studies, we just teach it over and over and over, never expecting them to master it at all. My favorite example of this is in mathematical topic of mean, median and mode. This topic is usually introduced somewhere around fourth grade and students see it year after year after that. Even Precalculus texts designed for 11th grade accelerated math students still include this topic. Eight years of learning and relearning a topic as basic as this is simply absurd. At what point do we stop this and say, "Enough! We expect you to know this now"?

This key principle was noted in the 2008 final report of the U.S. Department of Education's National Mathematics Advisory Panel[19]. The first of the reports main findings is as follows:

A focused, coherent progression of mathematics learning, with an emphasis on proficiency with key topics, should become the norm in elementary and middle school mathematics curricula. Any approach that continually revisits topics year after year without closure is to be avoided.

A good example of applying this in science is in the area of unit conversions. Beginning in Grammar School students should be expected to memorize common unit conversion factors such as 12 in = 1 ft; 5,280 ft = 1 mi; 60 s = 1 min; 365 days = 1 year; and others. In grades six, seven, and eight new factors such as 2.54 cm = 1 in and 1,000 cm^3 = 1 L can be added to the list along with common metric prefixes. (They should also learn which factors are exact and which are approximate.) By the time students reach ninth grade, there should be a standard list of prefixes, conversion factors and constants that students are expected to know. As students move into later studies in chemistry or physics new factors and constants can be added to this list in each new course. The list should be maintained and disseminated by the department, with the understanding that teachers at each level are to hold their students accountable for knowing and using the list appropriate to that level. Each new year teachers should distribute the

---

19 www.ed.gov/about/bdscomm/list/mathpanel/report/final-report.pdf

new list to their students, indicating which new items have been added, and informing them that they are still required to know and use all of the old information.

It should be obvious that this list of conversion factors, constants and prefixes, as well as many other particular points of scientific knowledge, function in science the way spelling lists, vocabulary lists and grammatical rules do in the humanities. No teacher of a foreign language could do her job if she had to start at the beginning each year, and every English teacher after first or second grade expects his students to have mastered certain points of grammar before entering his class. If we want our students to know anything in science we must establish the same expectations.

This point is important enough to merit stating it negatively in another example. If you expect students to possess certain knowledge coming into your class (and you should), then expect them to know it and do not put questions on tests that merely ask them to write it down. Instead, assume they still have and will use this prior knowledge in addressing new questions. If you teach chemistry and your students are supposed to know how to perform unit conversions when they enter your class, then do not put a problem on your test that states, "Convert 3.20 atmospheres into pascals." For sophomores or juniors in high school this is a trivial skill they should have mastered long before, even if they just learned what a pascal was that year. Instead, embed this conversion into a problem that is part of your course, such as calculations involving the Ideal Gas Law. If students are asked to use this law to solve for, say, a volume and the pressure is given in atmospheres, then they will have to perform the conversion to pascals as part of the problem.

Enabling Students to Succeed in a Mastery Environment

The techniques and tools we can use with our students to enable mastery and retention are virtually without number, and teachers who are constantly thinking about how to enable their students to succeed will think of new ideas all the time. In this section I am going to list many such ideas, all of which I have found to be very effective over many years. But before doing so I would like to stress that using these enabling strategies is, without question, critical and necessary for your school's success. To expect students to perform at a mastery level, remembering skills and facts they learned in previous months or previous courses, without the aid of an enabling support structure will simply not work. Every teacher knows this, because they know that if they are operating in a Cram-Pass-Forget environment and they simply start asking students questions on tests that come from old material, grades will plummet and many students will fail. So transitioning to a mastery environment at your school requires that you simultaneously implement as many enabling strategies as possible, that is, all of those listed below and more. Developing these strategies and the documents to support them takes a lot of time, but the satisfying result will be that your students actually do learn and remember what they have learned. So here they are.

Enabling Strategies

- *Use Objectives Lists to inform students about what they need to be able to do.*

  Students need to know what is expected of them. The more you expect of them, the greater the need to make it clear exactly what you want them to be able to do. Simply stating that we are having an exam "on Chapter 5" is inadequate. There are an infinite number of things you could ask students to do based on material from Chapter 5 in your text, some very easy (such as stating memorized definitions) and some very difficult (such as asking them to relate a concept to something you have never discussed in class).

  For each Chapter in your course, develop an Objectives List that states exactly what you expect your students to be able to do on quizzes or tests. State the objectives using action verbs such as calculate, define, explain, compare, distinguish, describe, and determine. Avoid immeasurable verbs such as know, understand, or be familiar with. In other words, use the same verbs on your Objectives List that you will use on your exams or quizzes. Below is an example Objectives List taken from my senior Physics Class:

Physics

Chapter 5: Work, Energy and Power

Objectives

The students will be able to do each of the following tasks, using supporting terms and principles as necessary:

1. Define and use correctly the following terms: work, gravitational potential energy, kinetic energy, elastic potential energy, and power.
2. State the Law of Conservation of Energy.
3. Describe what conservative and nonconservative forces are, and how they relate to conservative force fields and path independence.
4. Calculate work, gravitational potential energy, kinetic energy, elastic potential energy, and power.
5. Use the basic principle of the Law of Conservation of Energy, along with all of the specific types of energy studied in the Chapter, to solve application problems.
6. Develop solution strategies to multi-step problems that combine concepts from Chapters 1-5.

Give your students the reassurance that if they can do everything on the Objectives List they are ready for the test. Then stick to that commitment by not putting anything on a test that cannot be directly linked to the Objectives List. If you discover something you left off the list that needs to be on there, simply inform your class to add it in and make yourself a note to correct the list before next year.

Give copies of each new Objectives List to your students on the first day of the new unit under study. Then call attention to it frequently in class, informing students of topics you have already covered, and ones that are coming up. It is very helpful to students to be told things they are already supposed to know, particularly for students who were absent and have not yet followed up on what they missed. If you do this, then those who discover they don't know an item (for whatever reason) that has already been covered can come to you during your tutorial period and get the additional instruction they need.

Save the individual Chapter Objectives Lists into a single file and post this file on your class website. This way they are instantly available to all of the students in the class and their parents. Parents will feel informed, and students who lose their copy can easily obtain another one without coming to the teacher for it.

- *Orient your interaction with the class completely around mastery.*

Every single day in your class should be a mastery experience for your students. There should not be a single moment when they forget that they are in a mastery environment. As you lecture, direct frequent questions to students relating to old material. Refer to Objectives Lists frequently. Remind students often that their study regimen should be based heavily on review of old material and not just on memorizing new material the night before the quiz. Train your students to keep track of the kinds of questions they miss on your assessments so they know where to focus their review efforts.

If you use the weekly cumulative quiz routine I have described it is probable that unless you take preventive measures you will see student quiz averages begin to drop sometime in November. This will be because there is an increasing amount of material for students to review, and since they are new to a cumulative, mastery environment many of them will not be studying effectively to promote mastery. A wonderfully effective way to manage this is with weekly study guides (see below). But additionally, an effective and proactive intervention strategy to prevent this decline from setting in is to schedule regular face to face conferences with each of your students to discuss their study methods, expectations for the course, areas of difficulty, and so on. A couple of times per semester simply print off your roster with the grades and tell the students it is time for short conferences. Ask them each to come see you during a study hall, during tutorial times, after school, etc. Keep these conferences to just five minutes or so, and review the students' grades, how the student is studying, whether he is using the weekly study guides effectively, and whether he is putting in enough time. As each student comes by for the conference, mark his or her name off your roster, and make a point of going after each student who doesn't come as requested. This will demonstrate to each student that you care for them personally and that you are being conscientious not to let them fall through the cracks. Students who know they are not studying properly and who are in denial about it (hoping things will turn around by themselves on the next quiz without the requisite effort) will benefit by having you tell them directly what they need to be doing in their weekly study time and how they will succeed if they follow your directions.

Sometimes I have had students tell me quite plainly that they don't like to study and that it sure would be nice if they could make a good grade without all the effort required to master the material. In answer to this I always tell students equally plainly that I have designed the course specifically to give good grades to those who master the material and bad grades to those who don't, and that I have stacked the deck in the class so that they have no option but to master the material or fail. Contrary to the way it may sound, this is nearly always an enjoyable conversation when it comes up. We are both speaking candidly and the student knows quite well that I am right and that the reason he doesn't want to study for mastery is that it takes work. I usually end up quoting Hebrews 12:11 to the student and sending him off with a grin, a pat on the back, and an exhortation to start studying the way he knows he needs to. Students respect this approach.

- *Prepare weekly study guides.*

  This is a critical and extremely effective tool for students, and when I discovered how to use it in my own classes the results were dramatic. I use this technique in science classes in the pivotal ninth grade, and I think the same technique could be effectively used with sophomores. I am certain that younger students (seventh and eighth grade), who are still learning effective study skills, would profit from this technique.

  First, some background. I used to begin the school year by describing how to study for a cumulative quiz mastery environment. I had a detailed handout listing study strategies (flash cards, reviewing old quizzes, etc.) that I would pass out to the students. I would emphasize periodically how important weekly review of old material was and I posted a note to parents on the class website describing the philosophy of the course, its cumulative nature, and the importance of studying for retention. But despite years of effort at attempting to teach students how to study for retention, I found that this did not work for any one. For high achieving students, they took my warnings seriously and studied too much, and the vast amounts of time they spent was noted by the parents who reported back to the school that the demands of the class were too great. For other students, they did not know how to implement the strategies I gave them and they did not appreciate the importance of drilling with flash cards and other reviewing techniques, so these were neglected. As I've said, this really didn't work.

Here is what does work: Write a weekly study guide for the students. In the study guide describe what you would do that week to study for the class if you were a student in it yourself knowing that weekly assessments were cumulative. Distribute the study guide several days before the weekly quiz. Each week's guide can contain eight or ten specific study tasks, all of which can be accomplished in an hour or two. In the guide, tell students *how* to study in addition to telling them *what* to study. Instruct them to make flash cards for technical terms, equations, physical laws and so forth. Then, periodically remind them (in another edition of the weekly study guide) that new terms should have been added to the flash card set. Finally, tell them each week how many times to go through this flash card set (sometimes once, sometimes twice or three times). Have students set up similar flash card sets for historical information (see Chapter 10), and for unit conversions, unit prefixes, and physical constants you expect them to know. An example study guide from my ninth grade science class is shown below:

Accelerated Studies in Physics and Chemistry
Weekly Study Guide No. 11
Your assignments this week include the following review tasks:

1. Review last week's quiz and figure out what you lost points for. If you are not clear on how to improve your responses, visit with your instructor about it during the week. Do this before Friday.

2. Go through your five scientist cards once. Recite out loud to yourself everything you know about each of the scientists in the card set.
3. Rehearse the conversion factors, constants and metric prefixes.
4. Go through your technical term and equation cards twice. This stack is getting larger now and needs regular review.
5. Review the major types of variation and how to describe them properly.
6. Reread both sides of the Cycle of Scientific Enterprise handout.
7. Practice two problems for each equation we have learned so far. Be sure to review the calculations that have not come up as often on quizzes, such as weight or work.
8. Read over the previous Objectives List and make sure you can do all the things on it that we have covered so far. Raise a question in class if you are ever in doubt about anything.

In addition to instructions about reviewing flash card sets, list each week the other things you would do if you were in the class. You would know when it was time to reread a particular handout or summary sheet, when to review old notes, when to rework old practice problems, when to recite things out loud to yourself that you were supposed to know. You would also know to follow up on the previous week's quiz by looking up correct answers to things you missed, visiting with the teacher to revisit concepts you had not yet completely understood, and looking up the quiz keys to get correct answers so that you would have them available when reworking old quiz problems for practice.

All of these tasks should be completely spelled out in the weekly study guide, the contents of which will vary from week to week. When I began doing this the results were stunning. Students studied enough but not too much, quiz averages soared, and the answers I began getting, even to questions about material we had studied months before, were dramatically improved. When I ask my students how many of them use the study guides regularly and find them helpful, nearly every hand goes up and many heads begin nodding vigorously. For those few students whose grades stay low, a conference with parents that centers on how to use the weekly study guide should be enough for them to see that the class is accessible and the student has the tools he needs to succeed if he will put in the effort.

One final point for clarity: In the weekly study guide I do not "tip off" the students by directing them to study specific topics that will appear on the next quiz. This is not necessary, and it clearly subverts the entire philosophy of a mastery based program. But if the class expectations are gauged correctly and students are studying effectively by using the weekly study guide, real learning will occur and most students will be ready for almost any question at any time.

- *Use a variety of techniques to give your students adequate practice with skills both old and new.*

We teachers tend to take learning for granted, and we often forget just how hard it really is to master a new body of knowledge. But anyone who has tried to learn a new foreign language as an adult will encounter a set of challenges that make the issue crystal clear. My own studies in Latin (in which I am engaged, even as I write this) have been a refresher course in just how hard it is for students to learn and master new things. For example, in my Latin studies one week, I was working on exercises that included material that I had just done the previous week. I had worked hard on that older material, spending a diligent hour each day working through the exercises, but I had already forgotten it and was not ready to use it with the new material in the exercises at hand. That is, I had completed the homework, but I had not attained mastery. Sound familiar?

As I pondered this, two things struck me:

1. It takes a lot more practice to master something than is generally reflected in the length of the exercises in our textbooks. In mathematics, for example, one assignment of 20 or 30 problems to go with a lesson just isn't enough. It is enough for one night. But for most new academic skills, one night of practice is not enough time to achieve mastery. Neither are two nights or three. So we must routinely augment our problem sets with new problems revisiting old concepts and calculations. This is what mastery

requires, and most students will not seek after these additional exercises on their own. It's our job to create them and feed them to the students regularly.

2. There are at least three good ways to do this.

   a. Start up exercises: This technique is well known and is taught as standard practice throughout the teacher training literature. Almost every day when students walk into your classroom there should be a start up exercise on the board ready for them to work on. Your students should be trained that when the bell rings they are to stop visiting with their friends and start working on the start up. They will have only a couple of minutes to work on it while you are taking attendance, returning papers, and so on. As soon as you are ready to begin the lesson, begin by going over the solution to the start up and then move on to your lesson for the day.

   One good feature about the start ups is that students know these problems represent material they are already supposed to know. So we should tell them that if they ever have trouble with any start up exercises they should promptly come to tutoring times to get help. It's a good way for them to monitor their own level of preparedness.

   b. Write up new sets of exercises: Along with your daily homework assignment, occasionally write up new sets of problems revisiting old skills and pass these out too. I like to make these handouts relatively short (five or six problems) and mix up the problems types on the same sheet (five different problems from five different Chapters).

   c. Stop the car and get out: Instead of covering new material each day, or spending a day or two working exercises with new material, sometimes just stop and spend a day drilling on old material. This is especially helpful about two thirds of the way through a semester as final exams begin looming. You can help your students get a jump on exam preparation and feel better going into December or May by simply taking a day off now and then from marching through the curriculum and spending the day drilling old problems. I like to do this by flicking on my projector and typing the problem up for all to see. I work out the answer while the students do the same and I type it in as well. Then, if anyone had difficulty, we go over the solution. Then we do another one. By using the projector this way my problems are archived for next year. Students *love* this. For starters, the pressure is off because we aren't learning anything new. Second, they feel good getting renewed practice at something they know they are supposed to know because the truth is some students do get rusty on some skills despite my best efforts at promoting mastery. Third, each problem is a reminder of something the students are supposed to know. They like being reminded about something that has slipped their minds. Finally, the class period will be pleasantly paced because I am not trying to rush along to make sure I get through a lesson. We just work as many problems as we have time for and then we quit.

I don't consider these ideas as similar to options on a new car that you can take or leave, depending on your teaching style. They are more like the wheels on the car – essential to make it go.

- *Publish lists of prerequisite skills.*

  Prepare lists of skills students must master in one course and take with them as prerequisites in next year's course. This requires that very difficult thing—*cooperation* between teachers! It's all about establishing expectations, spanning over an entire department, that students are expected to live up to. As an example, students in my ninth grade science class, introductory physics and chemistry, go the following year into a course entitled Advanced Chemistry. Both of these courses are considered accelerated courses. The first several Chapters of the text used in Advanced Chemistry cover the same material that I covered in the last four months of the previous year. Rather than covering all this material over again, we inform students in ninth grade what they will be held accountable for knowing when they walk into the tenth grade chemistry course. The chemistry teacher's coverage of the first few Chapters is very condensed, and the course rapidly moves on to more advanced topics.

This cooperation entails the new teacher holding students accountable for prior learning, as I discussed in the previous major section. It also entails publishing to the students *what* they will be held accountable *for*. I have always been amazed at how many students will come up to the standard simply by being told exactly what the standard is. Many students, when I place the document in their hands telling them what they need to know on day one of chemistry, will make a point to review these things near the end of the summer to make sure they are ready! My class helps with this by requiring mastery of them in the first place so that they retain more of the knowledge and skills over the summer than they would have otherwise. Of course, many students don't do this, and thus they will have a rougher time of it for the first few weeks of chemistry. But even this experience will be another element of persuasion to the student that our program is about mastery, and the sooner the student starts studying with mastery in mind, the more successful he will be.

As another brief example, the increasing annual standards for scoring lab reports (Chapter 9) reflect this same mentality. Sophomores are expected to know how to write lab reports already!

- *For each science course publish a list of conversion factors, constants, and metric prefixes students are expected to know.*

I discussed this at length in the previous section. I bring it up again here merely to make the point that the conversion factor lists need to be *published* and *posted* on your course websites. Always strive to let students know explicitly what it is you expect them to master.

- *Offer troubleshooting for students whose grades get in a slump and won't pull up.*

  Occasionally a student's weekly quiz grade gets stuck in the 60s and even though the student seems to be studying diligently each week, the grade just doesn't come up. Over the years I have learned that this is usually a symptom of the student missing certain types of problems each time they occur. Once this is diagnosed by identifying the specific types of problems the student always misses the remedy is for the student to get additional instruction and practice on these specific areas. Once implemented this approach nearly always brings up the student's weekly quiz grade by 10 or 15 points.

  I developed a diagnostics chart to help with this. Across the top are all of the units we will study during the year. Down the left side are the specific types of items that occur on quizzes: computations, definitions or other memorized information, conceptual application questions, and historical questions. We then take out the stack of the student's quizzes and we go through them one by one marking the chart whenever we find an item on which the student lost a significant number of points. At the end of this exercise there are always a few boxes with several tally marks; these become the focus of our remediation efforts.

- *For assignments involving computations, make homework a truly useful learning tool by always giving the answers in advance.*

  By having the correct answers at hand students receive immediate feedback on whether or not they are doing the problem correctly, and thus the assignment will be as efficient as possible. If the students gets a correct answer, he knows he is on the right track and can keep moving with this confidence. If a student gets an incorrect answer he can stop immediately and try to find out why. This will save him the agony and pointlessness of doing an entire assignment incorrectly without knowing it.

  In fact, this is one of the key differences between an effective assignment and "busy work." When a student receives immediate signals marking her success or her error, she automatically begins striving to get the right answer on the first try. Motivation increases, and the student begins to pay close attention and to be very careful, knowing that these are necessary if she is to "win" by getting that right answer. This is, of course, exactly what we want the students to do. The parallels to contemporary electronic games are obvious, and it is no secret that students can exhibit terrific amounts of concentration at these. When students receive the answers week after week along with the assignment (I always print them on the back of the assignment handout) they rapidly learn that the point of the assignment is not to turn in the answers but to learn how to do the problems.

- *Compare notes with other teachers to find out what students are remembering and what they aren't.*

  If the students in seventh grade Life Science do not know that atoms are smaller than cells, we have a problem. The seventh grade teacher needs to confer with the Grammar School science teachers to find out if students were taught the hierarchical structure of matter (molecules are made of atoms, and cells are made of complex structures of molecules.) If they were not, then we have an issue with the Grammar School curriculum. If they were, then we have to begin inquiring into why students don't remember such a basic principle and how mastery of this might be improved in future years.

- *Make answer keys available so that old tests can be used as review tools for new tests.*

  One of the benefits of the weekly quiz idea for younger students discussed earlier is that each quiz becomes a new study tool for future quizzes. Students need to be encouraged to correct each of the problems they missed and rework them during a regular cycle of review of old material. Place copies of the keys for the quizzes, with solutions written out, in a binder in your classroom so that students have access to them during study hall and tutorial times. For computational problems, students who missed a particular problem can look up the correct answer so they will have it for reference during their review work.

Students who don't know how to work a problem can see the solution written out and spend time studying it.

One cautionary note is in order here. Some parents, once they learn that quiz keys are available for review, will insist that their child be allowed to photocopy the keys so that they have them at home as a reference. Since I reuse quizzes and quiz items from year to year, my policy has always been that students are allowed to review the keys as much as they wish, and may even make notes, but the keys are not allowed to leave the room and photocopying them is not permitted. Most parents understand the logic behind this policy and acquiesce without any problem, but be prepared for occasions when you have simply to insist.

- *Use the Daily Question to give students essential practice at forming answers to verbal questions.*

I described the details of the Daily Question routine back in Chapter 5 and I will not repeat that description here. However, in this section on enabling strategies it is essential to emphasize that for students to do well answering verbal questions they need practice at it. Telling 12- or 15-year-olds that they need to be able to explain the ways static electricity can form is one thing, but they need practice at it so that they can experience what it feels like to try to put it into words. They also need to be confronted directly with what a mediocre answer to a question looks like and what an excellent answer looks like.

- *Create a Standard Problems List for your class.*

This strategy applies particularly to students in upper division courses. At the beginning of the year in my Physics and Advanced Precalculus classes I give students a Standard Problems List that I prepared. This is a list of basic problems that appear in every text and that are commonly seen on standardized tests such as the College Board's SAT II subject tests. Students know that major exams will always include one or more problems or questions from older material represented on the Standard Problems List. This system helps students manage what they are supposed to know. If all they know is that a review problem can cover anything, they will likely be aghast or even in despair over the huge sweep of the teacher's expectations. But if you tell them review problems come straight off this list, they will be appreciative, encouraged and enabled.

Readers will have noticed that previously I described the weekly quiz regimen for a ninth grade class in which everything from the beginning of the year is fair game, but for upper division courses I suggest limiting review problems to topics itemized on the Standard Problems List. The difference is that the unit Objectives Lists for the younger students are much shorter (maybe 10 to 12 objectives) and the pace of the class is slower. It is not difficult for a diligent student to remember everything, and homework chiefly consists of reviewing. But in upper division courses the Objectives Lists tend to be quite long and the pace of the class is much faster. Material accumulates like snow from a snowplow and if students were on the hook for remembering every detail for every exam it would be too much. Additionally, because the problems and questions on exams in upper division classes are more complex, basic skills (such as vector addition) can easily be subsumed into more complex problems and do not need to appear as separate items on the Standard Problems List.

- *Be an advocate for your students.*

In addition to all of the pedagogical techniques I have described, you can have a huge impact on the ability of your students to meet your expectations for mastery simply by being an encouraging advocate for them. Some of us disciplinarian types are tempted to welcome the students who try hard and write off those who are indifferent. Years ago I thought such a strategy was warranted sometimes, but meditation on the way Jesus interacted with his often underperforming disciples convinced me it was not. Your students will be more inclined to work hard for you if they perceive that you are laboring on their side to help them succeed. This goes even for those students who are lazy or indifferent. Talk to them, go after them, counsel them, and maintain a positive attitude toward them. Let them know that it is your heart's desire that they succeed, and that you want to help them make that happen. Let their parents know the same thing. When the students see you pouring yourself out for them, every one of them, the effect on the overall performance of the class will be noticeable. Put another way, mastery expectations by themselves can seem harsh and intimidating. But when combined with an attitude on the teacher's part of loving and supportive advocacy the result is a perception of rigor with kindness. This is where we want to be.

PART III

Additional critical topics pertaining to your school's science program.

Chapter 8

Grammar School and the Science Curriculum

IN THIS CHAPTER I DO NOT INTEND TO PRESENT A LONG LIST OF SPECIFICS FOR science programs in the Grammar School grades. I will instead confine myself to a few specific observations relating to previous Chapters, plus one general recommendation about structuring the material for science studies in Grammar School. We will begin with some important cultural contextualization.

As everyone who works with children knows, young children are fascinated by the physical world. The only things one has to do to stimulate the interest of a four year old in animals like turtles and squirrels is to expose the child to the animal and display interest one's self. In a previous era the last phrase of the preceding sentence would perhaps not have been necessary, but in the age of video and text messaging it is.

Two fundamental things have changed in the past 50 years that relate to the issue of children developing an interest in the natural world. First, children do not necessarily grow up playing in the woods and around creeks the way they did two generations ago. They spend vast amounts of time in front of computers and televisions, activities so captivating, what with digital graphics and so on, that children can easily find stepping into the back yard or the field next door boring by comparison. Second, parents and others in the child's community can be so busy with their own technologies—mobile phones, DVDs, computers—and with their business and social lives, that they omit to display any interest in nature as part of their everyday lives. Even conscientious parents can make this mistake. Many are so busy that when they do have time to spend with their children they spend it reading with them, helping them with their math, or engaging with them in a variety of leisure activities and sports, rather than down at the local creek scooping up tadpoles. These activities are all laudable. But my point is that the combination of distracting, high-tech toys and lack of parental modeling make it easy for a child to grow up these days isolated from the fascination with nature that was the background for many of us when we were children.

One would think that the high-tech world we live in would be a new stimulus to children that would spark an interest in science. But ironically, for many (or most) children it actually interferes with their natural interest in nature. The technology in an mp3 player or mobile phone is so advanced that no youngster can possibly understand it, and thus these devices become fascinating "black boxes" that can perform fabulous tricks, the internal workings of which are an utter and impenetrable mystery. The bottom line for all these effects is that it is easy now for children to be either bored by nature or jaded by high-tech products, and without significant influence from parents and teachers to stimulate an interest in nature.

There are a few children who are exceptions to the trend toward jadedness, the so-called "computer geeks." As we all know, these interesting people are naturally inclined toward science and any stimulus at all in this direction typically unleashes a torrent of interest that sends the kid off to the internet to read far more about things than we would ever dream of requiring the general student population to learn. These are not the students we have to target as we consider enabling children to learn the scientific disciplines; they tend to take care of themselves, and end up being leaders in science classes, helping other students to understand what comes easily to them (and often discovering and gaining confidence in their own identities by doing so).

The painful irony here for us science teachers is that life in the 21st century demands scientific literacy as never before, but the huge technical advances that are the fruits of that literacy are interfering with the very literacy that we now need so badly. Thus, it is right for faculty and administration at the Grammar School level to be very interested in stimulating the interest in the natural world that may otherwise remain dormant. Consequently, getting kids outside to explore the natural world for themselves needs to be a major goal of the Grammar School science program.

Basic Principles for Grammar School Science Instruction

So leaving the details to others, here are a few basic principles that should be incorporated into the Grammar School science curriculum.

- *Introduce students early to appropriate ways to think and speak about science.*

  Third grade is not too early to introduce students to a good working definition of what science is, and to the distinction between truth and scientific facts. The Cycle of Scientific Enterprise should be presented in fourth grade. Students at this level should already be developing appropriate language for speaking about scientific theories and hypotheses. As we saw in Chapter 4, confusion about the most essential and basic terms involved in scientific inquiry, theory and hypothesis, is ubiquitous, and we should endeavor, even in these early years, to help children develop an appropriate vocabulary for talking about how science works. These concepts should be revisited and emphasized every year in scientific studies so that by the time students reach high school they are already thinking appropriately about scientific knowledge. Further, by introducing the Cycle of Scientific Enterprise to students in Grammar School, it can serve as a topic for further exploration in the middle school years, when studies in English composition or Logic are focusing on the development of logical thought, use of accurate definitions and unambiguous expression.

- *Integrate studies in science with those in other disciplines.*

    It is proper for Grammar Schools to focus on the three Rs, and when we do so, it can be difficult to work a meaningful allotment of time into the curriculum for science. Combining explorations in science with studies in other disciplines is an excellent way to make the curriculum efficient.

    Writing observations from nature in a science journal can be connected to poetry (most descriptive word choice). Essays and stories written for English can be based on explorations from science and will help students develop skills essential in scientific writing (proper syntax for clear expression, unambiguous word choice). Musical studies can be enhanced through the use of home-made instruments that facilitate inquiry into the ratios and proportions upon which musical scales are based. Historical studies can be keyed to advancements in technology (easily done with any study on a major war, or on cultural shifts due to available modes of transport or communication).

- *Set up a science and nature center.*

    Our school is blessed with a stupendous Science and Nature Center that is equal parts garden, farm, and park. Such a center is an outstanding asset for any Grammar School program. Students can plant seeds, raise the plants, and even consume the vegetables. Farm animals are not only interesting and fun to study, but can also provide students with periodic opportunities to welcome a cute new colt or kid into the world. Milk, eggs, and goat cheese can all come from this part of the center. Development of a nice pond can be taken as far as the school's resources will allow. A pond project can result in a complex ecosystem that will support studies in life science, as well as provide the school community with a haven for rest and leisure. Outbuildings or pavilions at the Center can provide venues for lessons or group activities, or function as natural history centers, housing collections of rocks, snake skins, fossils, and other items found in the vicinity. A greenhouse will provide further opportunities for research and projects.

    The science and nature center could be a key site for studying rainwater quality, collection, and conservation; solar power technologies; organic methods of pest control; and responsible land use. Regional water supplies and their use and preservation will be topics that arise naturally in the nature center. Regional geology and its relationship to human communities and urban development can also be highlighted. The possibilities are endless.

The development of a science and nature center is a great venue for parental involvement. Parents can take a lead role in developing the center, coordinating activities there among the students, and communicating events to the wider school community, knowing that their labors will directly support the education of their own young children. Agricultural clubs such as 4-H can also make arrangements with the school to share facilities so that maintenance is shared and everyone benefits.

One of the most ambitious ideas I have heard would be for students to assist in growing vegetables that are harvested and cooked in meals in the school cafeteria. Students could work together with employed staff to produce nutritious organic foods, and enjoy the fun of eating their own produce for lunch. There are great opportunities here for an integrated program that results in scientific study, inexpensive food, and nutrition.

Of course, many schools do not have the land to support a nature center and need to consider other options for regularly bringing hands-on nature studies to their students. Perhaps one of the school's families or friends owns land nearby that the school could use, or perhaps a nearby business has some extra property and would support the school by allowing them to use it. Schools in dense urban areas may not have access to any of these options, but there is a thriving movement now in urban farming that urban campuses can tap-in to. Perform an internet search on urban farms and you will find hundreds of examples (millions, actually) of urban farming on plots as small as a few square feet, or utilizing above-ground containers. Small greenhouses will enable farming activities to continue year round. These little farms can be used for growing vegetables, flowers, and various small animals. Even schools with a campus in the heart of a city and entirely paved can implement some of these innovative urban farming ideas.

Curriculum Content

When it comes to outlining the actual content the Grammar School science program will address, I believe it is critical that the curriculum distinguish between knowledge, skills and experiences. To this end, I have compiled the chart on page 112, descriptively (and perhaps whimsically) entitled, "Things Rising Seventh Graders Should Know" (TRSGSK). This chart outlines particulars to be addressed in each of these three domains.

The first section of TRSGSK summarizes knowledge content that can be distributed throughout science classes in the Grammar School grades. It is quite reasonable to expect that students who have completed sixth grade would have a working understanding of each of the topics in this chart. A few important points apply here.

First, make these explorations as hands-on as possible. Depending on the school's resources, endeavor not to take up any of these topics without providing students opportunities to see, hear and touch the real thing for themselves. Even topics such as organ systems should not be studied merely from a book. There are plenty of hunters around who dress their own meat, who would feel it an honor and privilege to be asked to preserve the organs for show and tell at the school. This may sound messy, but imagine the impact on the boys in the room when one of their dads brings in actual deer organs for students to look at! In fact, I would submit that at this level text books are of little value compared to the impact of a first hand encounter, supported by the student's own documentation of the event in his science journal.

Second, require students to develop and use appropriate technical vocabulary as much as possible. Do not underestimate students' ability during these formative years to learn and use an extensive scientific vocabulary. Although adults often shy away from the massive number of technical terms that arises when studying this myriad of topics, students will love them (if they are well presented by an enthusiastic teacher) and will take pride in the sense of ownership they feel when they can explain things to their parents using words the parents don't even know.

Regarding the development of technical vocabulary, a word of caution is in order here. Teachers should avoid the temptation to teach the vocabulary, administer a vocabulary quiz and then consider that they have done all that is needed. Engaging students in an interesting science study entails a lot more than merely learning vocabulary. Instead, teach the vocabulary, use the terms in class discussions, and encourage your students to use proper vocabulary in class. Then on your quizzes and tests write items that require students to know and use the vocabulary in order to understand the questions and form acceptable answers to them.

The second section in TRSGSK itemizes skills students should develop, at least at an introductory level, before they leave the Grammar School. The only remark I need to make here is to state the obvious: Students learn how to do these things by doing them, over and over and over. If energetic teenage boys arrive in the high school without ever having learned how to act in a laboratory or how to use care when handling delicate specimens, there will be trouble. Conversely, anyone can learn appropriate behavior as a child and bring this training into a new context that comes along later. Again, time spent documenting observations in a journal, and being corrected when these observations omit important detail or are awkwardly worded, will directly enhance linguistic development and will equip students to use lab journals effectively in later studies.

The third section of TRSGSK contains a list of experiences every Grammar School student in our country should have. Most of these are activities that are simple to organize, such as purchasing inexpensive magnets and a bottle of iron filings and letting students play with them. Others can be accomplished during school trips, such as visits to caves or to natural environments that are different than those in the immediate vicinity of the school. Many of these experiences are things adults tend to take for granted. We have seen so many radiographs (the appropriate technical term for the images commonly referred to as x-rays) that we don't think anything about them. But if an eight-year old child gets a chance to look at an image of the bones of someone he knows, the experience will be unforgettable and furnishes an excellent opportunity to learn the names of the bones in the image. If one of the students in the class breaks an arm and the class doesn't get to examine the radiograph in detail a premium educational opportunity is lost. Better still: If the child's doctor is a friend of the school, invite the doctor to come in and explain the radiograph, complete with correct terminology!

Things Rising Seventh Graders Should Know (TRSGSK)

Knowledge

General	Physics	Chemistry	Biology/Physiology	Earth/Cosmology
Common U.S. and metric system units of measure	Atoms, protons, neutrons, electrons, electric charge	Elements, molecules, compounds	Types of animals and their habitats	Weather, clouds, hurricanes, tornadoes, thunder, lightning, water cycle
Scientific Method	Laws of Motion	Three basic phases of matter	Classification of organisms	Ecosystems, greenhouse effect, climate change, erosion
Benefits and consequences of technology	Magnetic poles; opposites attract; magnetic fields	Chemical reactions	Cells/molecules	Earth's magnetic field, compasses
Distinction between truth and scientific facts	Gravity; distinction between mass and weight	Combustion	Organ systems and function	Types of rocks
Science as model building	Density	Chemical and physical changes	Organ structural hierarchy: cell–tissue–organ	Pollution, recycling
Relation between facts, theories, hypotheses, and experiments	Heat and temperature; F and C scales	Chemical and physical properties	Human diseases, vaccinations, hygiene, microorganisms, viruses	Volcanoes, earthquakes, tsunamis
	Energy, forms/conversions of energy	Properties of water	Metamorphosis	Moon, moon phases, tides
	Waves	Solutions, concentration, dissolving	Heredity; genes	Oceans
	Sound and music	Names of common lab apparatus (beaker, flask, test tube, etc.)	Effects of substance abuse	Solar system, astronomy
	Light, color, prisms, lenses, lasers			Fossil fuels and greenhouse gases
	Simple machines			Dinosaurs, fossils, geological formations
	Electricity			

Skills

- Writing accurate and detailed observations in a journal.
- Describing procedures, processes and results using clear, technical language.
- Making measurements with analog instruments, e.g., scales, thermometers, meter sticks, measuring cups, graduated cylinders.
- Identifying units of measure when recording measurements.
- Identifying properties and characteristics of substances.
- Using appropriate care when using and handling apparatus, samples, and organisms.
- Following safe practices and exhibiting appropriate behavior in a laboratory.

Experiences

- Perform controlled experiments according to the steps in the Scientific Method.
- Make first-hand contact with growing plants and caring for animals.
- Watch animals being born.
- Look at cells, bacteria, snowflakes and other objects with a microscope.
- Make solutions.
- Grow crystals.
- Watch chemical reactions; discuss properties of reactants and products.
- Work as part of a team on science projects and experiments.
- Watch videos of weightless astronauts.
- Burn things with a magnifying glass.
- Make a siphon.
- Play with magnets and iron filings.
- Look at the moon and planets through a telescope.
- Examine minerals and rocks.
- Create melodies with home-made instruments, such as a set of bottles filled to different levels with water.
- Look at radiographs (aka x-rays) and sonograms.
- Visit a cave with its stalagmites, stalactites, etc.
- Spend time in or near the following natural environments: forests, deserts, oceans, ponds, snow, and rivers.
- See and hear a lightning storm.
- See videos of tornadoes and hurricanes and the destruction they cause.
- Watch changing tides.
- Use compasses and maps.
- Look at the colors of light refracted through a prism.
- Experiment with batteries and simple circuits.
- See lasers used in various optical applications.

Chapter 9

Laboratory Work and Lab Reports

When I described the Core Principles for Instruction in Science in Chapter 1, the third principle about the tools of scientific thinking included a number of items relating to laboratory work. In this Chapter I mainly wish to address the way lab reports relate to our curricular objectives for science, but a word about laboratory work in general is also in order.

There are two extremes to avoid when considering the laboratory component in a high school science course. The first is to assume that since a school has limited resources, or is "classical," laboratory work can be neglected or ignored. Laboratory work is indispensable in every serious course of scientific study. The other extreme is to insist, as some public school boards do, that laboratory work is so critical that 40% of class time in all science classes must be devoted to it. Spending this much time in lab work will mean that students do not have the classroom exposure time necessary to assure that they master crucial concepts, computations and vocabulary. It also means that lab activities become superficial and trivialized, because teachers do not have enough time to develop enough meaningful lab explorations to occupy two out of every five days of school.

I have taught at schools with massive budgets and at schools with severely limited budgets. Although having plenty of money allows science teachers to purchase lots of toys, it is also the case that many explorations in physics and chemistry can be pursued with very limited resources without compromising the value of the laboratory work in the classroom. In fact, many of my favorite laboratory explorations in physics use home made equipment or common consumer materials. It is true that to develop a sophisticated program in molecular biology, an appropriate accelerated program of study for juniors and seniors, will require a significant outlay for equipment and facilities. But such is not the case across the board.

Schools that call themselves "classical" may be operating under the mistaken notion that a classical school science program is one that has students reading Aristotle's *Physics* rather than engaging in laboratory work. Nothing could be further from the truth. "Classical" does not equate to dusty and impractical. In fact, as I described in the Introduction, the whole idea of a classical education is centered on leading students to think critically and integratively in whatever arena they find themselves, and in contemporary society comprehension of scientific principles cannot be adequately developed apart from an effective program of laboratory work.

On the other hand, many canned experiments, and the specialized apparatus marketed by science equipment suppliers to support them, can only be described as fatuous. Experienced science teachers know this, but new science teachers will do well to give serious consideration to the actual value brought to the curriculum by every laboratory investigation, and not to assume that a particular lab activity should be included in the curriculum simply because other schools include it.

A valuable lesson students can only learn in the context of laboratory investigations is that scientific research is not about getting the exact answer one expects. There is always experimental error due to limitations in equipment and experimental technique. The question is how much error a result can have and still be considered a satisfactory result. This simple idea is second nature to practicing scientists, but is lost on students who have not been taught to think in these terms. Laboratory investigations are the indispensable forum for developing an understanding of this principle. Accordingly, a major focus for student lab reports should be assessing experimental results, both quantitatively and qualitatively. Students should compare their results to the original hypothesis and determine if the hypothesis has been confirmed. They should compute experimental error and judge whether the error is reasonable, given the equipment and procedures used. And they should also judge whether the experimental results were definitive, regardless of the amount of error or whether or not the hypothesis was confirmed.

Learning Objectives for Upper School Science Labs

When planning any course of instruction it is essential to identify specific objectives for student achievement. Much has been written about how to write learning objectives effectively, and it is certainly possible to get carried away trying to specify exactly what students are supposed to be able to do, under what conditions, and so on. The bottom line is that for any course of instruction the teacher should specify to students what it is that she expects them to be able to do as a result of the study, and the teacher's lessons should be designed to accomplish this. It is also the case that the teacher's lessons will be better structured, more efficient, and more likely to accomplish their purpose if learning objectives for the lessons have been specified in some detail.

The same holds true for laboratory investigations. Lab activities need to be far more than just activities that bring diversity and interest into the science program, or enhancement of material learned in lecture. There are specific elements of science knowledge that lab activities can very effectively address. With this in mind, I propose the following set of objectives to be used for laboratory activities:

Learning Objectives for Science Labs

After successfully completing the required science courses at this high school students will:

1. State and follow standard laboratory safety practices.
2. Correctly identify and use standard laboratory apparatus.
3. Use proper care in setting up apparatus to protect equipment and minimize experimental error.

4. Describe and follow the proper methods for making measurements with common instruments. This includes identifying the types of errors that can introduce inaccuracies in measurements and describing how to avoid them.
5. State the role of precision in taking measurements, and relate this to the significant figures in a measurement.
6. Apply the Scientific Method to conducting experiments and to writing reports.
7. Apply appropriate logic to conducting an experiment and to writing the report.
8. Maintain a proper lab journal.
9. Clearly explain the theoretical background behind an experiment using quantitative analysis where appropriate.
10. Use quantitative predictions from scientific theory to form testable hypotheses.
11. Clearly and efficiently describe a scientific procedure and the results and discoveries that followed.
12. Use appropriate care in experimental procedures and data collection.
13. Present calculations and data in a clear, organized fashion such that others can verify calculations or check results. This includes development of tables and graphs using standard scientific units and formatting.
14. Apply quantitative error analysis to experimental data as appropriate.
15. Apply qualitative analysis to experimental results as appropriate.
16. Estimate uncertainty in measurements.
17. Apply cogent reasoning to analysis and discussion of experimental results. This includes reasonable identification of the factors which contributed to experimental error.
18. Use computer tools to take data, graph data, and develop reports.
19. Use clear, concise, and accurate language in a technical style in scientific reports.
20. Cooperate with team members successfully to accomplish each of the above objectives.

Hand Crafted Lab Reports

In Chapter 6 I broached the issue of abandoning the use of canned "lab manuals" in favor of having students write their own lab reports from scratch. Prior to entering high school many students would not have the verbal and technical skills necessary to tackle such an assignment. But as they enter high school they should, so I write now of students in grades 9-12. In Chapter 6 I outlined several of the disadvantages associated with using canned lab manuals for laboratory activities. There are also a great number of positive advantages to having students write their own reports. To write good lab reports students must:

- Journal experimental procedures and findings so that they have documented evidence from which to construct the report.
- Use standard English in a technical style to communicate scientific material.
- Spell technical terms correctly.
- Identify the purpose for conducting the experiment and the particular hypothesis under test.
- Contextualize the experiment by writing a background section that frames the problem (the big picture), describes the necessary theoretical background, and sets forth the general experimental approach that was followed in the exercise.
- Name all of the materials and apparatus used in the experiment.
- Thoroughly explain, in a clear and logical fashion, the procedure followed in the experiment.
- Evaluate the experimental results, qualitatively and quantitatively, and discuss whether the results confirm the hypothesis or not. Or, in the case of a descriptive laboratory investigation, develop a thorough and accurate description of what was discovered.
- Determine an effective form for presenting experimental data, including tables, charts and graphs, and use computer tools to develop these presentations and incorporate them into the report.
- Make decisions about a host of non-trivial formatting issues such as table layouts, methods for displaying units of measure, titles for tables and graphs, and scales and labels for axes on graphs, all of which either enhance or enable the reader's comprehension of the experimental work.

- Assess the experimental procedure, apparatus and method and evaluate the adequacy of the results and the overall success of the experiment.

Obviously, these skills relate directly to virtually any school's overall educational mission, so I believe further justification for teaching students to write their own reports is unnecessary. However, it is also the case that many students will be able to use these skills directly in their endeavors after high school. I have received a number of testimonials from current or former students indicating that employers or instructors in collegiate programs were very impressed that our students already knew how to use a lab journal and write a standard report.

Lab Report Handbook

Learning to write lab reports is a significant challenge. To manage it well students will need comprehensive resources they can refer to for guidance on style, content, formatting, punctuation, and instructions for using computer tools. They will need to see examples of comparable reports, and will need detailed descriptions of what to include and what to avoid in an effective report. Instruction should be based on standard formatting and style used in scientific journals. I spent several years developing various handouts and guidelines for my students before finally compiling them into *The Student Lab Report Handbook* that our students have been using for the past several years.[20]

Lab Report Proficiency Requirements by Year

If the humanities faculty at your school have done a good job structuring and defining the writing curriculum, they will have specified skills students are expected to demonstrate and master during each year in the program. Expectations and grading standards among faculty will be consistent, and will increase consistently year by year with the students' increasing maturity and experience. The same should apply to writing lab reports in the science

---

20 *The Student Lab Report Handbook* is available from Novare Science and Math at novarescienceandmath.com.

program. Expectations and grading standards for lab reports need to be consistently established across the department. Each faculty member must do his or her part to uphold the standards, instructing and grading students fairly and rigorously according to a rubric that is consistent with the science department's objectives for lab activities.

Beginning in ninth grade students should conduct in-depth lab investigations and write a full report of the work at least five or six times per year. The goal for student reports, which should be announced to them when they begin writing these reports in ninth grade, is that they will have mastered the skill of writing lab reports by the end of the fall term during their sophomore year. This gives students a year and a half, with at least eight or ten separate report writing experiences, to master this crucial aspect of the science program. After that, students can devote their attention to conducting and documenting the laboratory investigation, without also having to figure out how to write a report.

Below is a chart outlining grading criteria that may be applied to science courses in grades nine through twelve.

Lab Report Grading Criteria and Writing Objectives

Freshman Courses	Content Objectives	To earn an A	To earn a B	To earn a C	To earn an F
First Report	1. Correct content and adequate descriptions. 2. Data presentation. 3. Correct English grammar and punctuation. 4. Typed with proper formatting. 5. Citations for historical references.	An excellent first attempt. Paper is thoroughly marked for future improvements.	Good work but much improvement needed. Student made a very good effort to meet requirements but significant improvement is needed in key respects. Paper is marked thoroughly.	Student made a good faith effort to meet requirements but the report contains serious deficiencies. Paper is marked thoroughly.	Good faith effort to meet lab report requirements is not evident. Teacher conference with student is in order.
Second Report	1. Error calculation and discussion. 2. Evidence of care in proofing, editing, formatting. 3. Improved technical English style; accurate use of technical vocabulary. 4. Improved Background, Procedure and Discussion sections. 5. Graphs prepared on Microsoft Excel. Graphs may include experimental and predicted curves (especially for accelerated classes).				
Third Report	1. Graphs prepared on Excel, with appropriate fonts and labels. 2. Evidence of care in experimental procedure and data collection. 3. Most if not all formatting requirements are met. 4. Quantitative error discussion.				
Fourth Report	1. Graphs are virtually perfect. 2. All formatting requirements are met. 3. Clear, accurate use of technical vocabulary.				

(continued)

Lab Report Grading Criteria and Writing Objectives (continued)

Sophomore Courses	Content Objectives	To earn an A	To earn a B	To earn a C	To earn an F
	1. Graphs include predicted (theoretical) and actual curves. 2. Error discussion is thoroughly quantitative and relates directly to differences between theoretical and actual curves. 3. By midyear, reports should consistently display correct formatting; clear, accurate technical language; and good discussions of error.	An excellent paper, virtually perfect, which evidences good laboratory procedures and care. The student is clearly building on report writing skills learned the previous year.	Slight imperfections such as deficient style, unclear descriptions or incorrect formatting, but still a very good effort.	Content is all there and adequate; descriptions are serviceable but not great; numerous deficiencies in formatting.	Evidence is lacking that the student has already spent an entire year learning how to write reports and use care in experimental procedure.
Junior/ Senior Courses	Content Objectives	To earn an A	To earn a B	To earn a C	To earn an F
	1. Reports should include abstracts. 2. More sophisticated data analysis is included, using tools in Microsoft Excel. 3. Graphs include error bars (based on analysis of σ, etc).	An excellent paper, virtually perfect; thorough display that student uses solid laboratory methods and care, and has learned how to write and format a solid report.	Slight imperfections such as deficient style, unclear descriptions or incorrect formatting, but still a very good effort; error discussion is thorough and grounded in quantitative analysis.	Content is all there and adequate; descriptions are serviceable; some deficiencies in formatting or analysis.	Evidence is lacking that the student has already spent two years learning how to write reports.

Chapter 10

Making History Work in Your High School Science Class

Every school seeking to build a balanced liberal arts curriculum (which includes the humanities and the sciences) will acknowledge the value of teaching students key historical figures and events in the history of science. As I mentioned back in Chapter 1, one of the core principles of science pedagogy is that the history of science helps us understand how scientific knowledge works (the Cycle of Scientific Enterprise), and thus facilitates understanding science itself. But incorporating historical lessons effectively into upper school science classes is challenging. How does one encourage students to know major names, events and discoveries in the history of science without simply handing them a big list of historical facts to memorize? How does one insure that students are able to recall the most important facts from a one-hour history lesson? Even more importantly, how does one get them to remember things rather than simply forgetting them after the test? If students forget who Antoine Lavoisier was three weeks after the test, then why even go to the trouble of mentioning him in the first place?

Richard Feynman, a Nobel Prize winning physicist, once said that "history is fundamentally irrelevant."[21] However, he was referring specifically to the creative insights needed for new theories on the frontiers of physics. As educators who promote a curriculum based on liberal arts, we affirm the importance of history (Feynman did too, and knew his scientific history), the integration of all knowledge, and the significance of cultural literacy. Accordingly, in this Chapter we will explore how to make history a vital part of your science course without making it a burden.

Start with careful selection of historical material. Historical connections should include names and nationalities of scientists, key dates of new discoveries or publication of new theories, and key features of historical theories. But the material chosen for inclusion in your course should be immediately relevant and directly related to the content at hand. Make sure the material is worth the trouble of learning by focusing on scientists of real significance, not just on your particular favorite scientists or time periods of interest. Put clear limits on the amount of material you expect students to learn. Detailed study of scientific treatises from the ancient classical era, or even from the Renaissance, is neither feasible nor desirable in a contemporary science course, nor does such study make science "classical."

Second, clear packaging of student expectations is crucial. Develop a list of scientists for your course that students will be expected to know about and distribute this list, together with the description of what students need to know, at the beginning of the school year. I have placed an example of such a document at the end of this Chapter. In it I have divided scientists into three categories of significance, with a descending scale of what students need to know and how much the knowledge is worth on quizzes. For Category I scientists (like Galileo), students need to be able to write a one-paragraph biography summarizing major works and life events. Students also need to know the scientist's name, nationality, one or two key dates (years), and a brief description of the scientist's most famous experiment or contribution (such as Rutherford's gold foil experiment or Dalton's five point atomic model). Questions about scientists in this category can be worth 10 to 20 points on quizzes. Category II scientists require

---

21 *Richard Feynman: Take the World from Another Point of View*, Films for the Humanities and Sciences

knowledge of name, nationality, a key date or two, and a three or four sentence summary of the scientist's major scientific discovery or experiment. Questions about these scientists are limited to a maximum of 10 points in value. For Category III scientists students need to be able to write a single sentence that includes the scientist's nationality, major contribution and the year it was published. Quiz items on these scientists are limited to five points in value.

The entire list of scientists for one science course should be limited to about a dozen names. Students will encounter these names in clusters during the year as new topics are introduced, so that learning about these historical figures is not burdensome. Keeping the knowledge fresh will literally require only a few minutes each week to go over the flash cards the student has made (see Chapter 7). We all know that several years later the students may have forgotten many of the details about these scientists, but after rehearsing them for an entire year they will also retain a lot, and their ability to relate quickly to these names when they come up in future studies will be virtually permanent. I have frequently had seniors in my physics class who remember a significant portion of the historical information they first encountered when they were in my ninth grade science class.

Finally, make your presentation of the history engaging. This is, of course, simply good pedagogy, but it can make a huge difference to the dynamics of your class if the students are eager to encounter new historical material because they know it will be fun. In recent years I have developed PowerPoint slide shows to use when introducing the history behind new units of study. I pack these shows with images, .gif files, diagrams, key historical details, dates, and interesting side notes. For continuity and interest, my presentations include historical figures that are not on the list of names the students need to know about. Naturally, the students don't mind this at all.

I try to make the shows as whimsical as possible so that students laugh frequently. Who can forget Archimedes running naked through the streets of Syracuse shouting, "Eureka!"? What a drama was played out when Galileo went to trial! What pathos there is in the story of the great Michael Faraday getting his humble start as a bottle washer in Sir Davy's lab! How about those twitching frog legs that led Count Volta to discover the electrochemical cell, the forerunner of the modern battery? Or Tycho Brahe's artificial nose, made necessary because he lost parts of his nose in two separate duels? Studying the history of science is great fun, and knowing the exploits of the scientific heroes of the past properly humanizes a discipline that can become sterile when divorced from the human beings who are doing the science.

Tycho is one of my favorites. You should have your own favorite (but relevant) scientists and talk about them as if they were your good friends. To make your presentations authoritative as well as interesting, you will need to do your reading. One of my favorite resources is George Gamow's engaging history, *The Great Physicists from Galileo to Einstein*, which, despite the title, goes back to the ancient Greeks. This delightful book is packed with funny anecdotes and solid historical information, such as Galileo's incredible recantation speech, which I enjoy reading to my students every year.

After showing a slide show to my students I place it on the class website so they can review it at their leisure and prepare their flashcards. All of the information they need to be able to answer the questions on the weekly quizzes is included in the slides. They love this because it means they don't have to take notes during the presentation (though many do anyway). This goes a long way toward making it fun. It also doesn't hurt to be a bit of a showman with a penchant for hamming it up during the presentation!

As mentioned above, what follows is a sample of handout that can be given to students during the first week of school. The handout apprises them of which scientists they need to know about and exactly what they need to know. The scientists listed in this example are the particular ones I chose for emphasis in one of my own courses. There are many other important scientists that could be added, but the list needs to be limited to about a dozen names for one course.

Scientists Required for Cumulative Review

During the school year we will be studying a number of histories in scientific thought in the West, and with each of them there will be a number of scientists we will encounter. Some of these contributors are of such key importance that they are worth remembering. This sheet will help you organize your study of these key scientists.

These are the names of the scientists you will be expected to know about on quizzes this year. The two tables below indicate three categories of scientists for our studies and the things you need to know about the scientists in each category.

What to Know	
Category 1 (10 to 20 point quiz items)	• His name and nationality • Key dates in his life • Where he worked • A one-paragraph summary of his major discoveries and contributions and key events in his life
Category 2 (10 point quiz items)	• His name and nationality • A one-paragraph summary of his major contribution and the year it was published
Category 3 (5 point quiz items)	• His name and nationality • A single statement describing his major discovery, contribution, or well-known experiment, including the year

Scientists			
Topic	Category 1	Category 2	Category 3
Medieval Cosmology and the Copernican Revolution	Tycho Brahe Johannes Kepler Galileo	Nicolas Copernicus	
Motion/Gravity	Isaac Newton	Albert Einstein	
Electricity/Magnetism			Alessandro Volta Luigi Galvani
Atomic Models	John Dalton	J. J. Thomson Ernest Rutherford	
Chemistry	Antoine Lavoisier		Dmitri Mendeleev

Chapter 11
Dealing with Evolution

THUS FAR THIS BOOK HAS BEEN ABOUT THE PRINCIPLES AND MASTERY OF SCIENTIFIC knowledge and how to develop them in the classroom. I have not discussed methods or principles pertaining to specific content areas. But given the nature of the ongoing controversy over evolution among Christians in our country, I would be remiss to publish a book on science education for Christian schools without devoting some attention to this contentious issue. I suppose there are Christian schools that accept evolutionary theory and teach it, without reservation, as the way things are. My school is not one of them and, of course, there are thousands of conservative schools who would not dream of taking such a position. In this Chapter I am not going to argue for or against the theory or its relationship to interpreting Genesis. But I do want to set forth what I believe is the responsible way for Christian schools to address the issue, an approach that should cause no offense to those who find evolution incompatible with Scripture, but which is still satisfying to Christians who fully embrace evolutionary theory.

Christian schools have a very important responsibility with respect to this topic. In our day we are witnessing an historical conjunction of several significant factors:

1. Scientific knowledge is regarded as paramount by the culture at large.

2. Genetic research influencing the current state of evolutionary theory is constantly in the news.
3. Conservative Christians, in general, continue to oppose evolutionary theory, and are often involved in highly publicized and controversial actions by school boards and other groups.
4. The highly publicized dispute between conservative Christians and the majority of the scientific community continues to rage.
5. It is routine to hear of Christian students who go off to college and are shocked by the claims of those promoting evolutionary theory, and who enter into a crisis of faith as a result.

It is imperative that those of us striving to train our students well and to equip them to live "purposefully and intelligently in the service of God and man," as my own school's mission statement puts it, embrace a rigorous and thoughtful approach to dealing with this subject.

One of the primary elements fueling the controversy is the widespread misuse of the term "theory." Everyone misuses this word these days, Christian and non-Christian alike. I constantly hear people referring to evolutionary theory with phrases such as, "Oh you don't have to believe in that. It's just a theory." It is simply incorrect to speak of theories in this way and our efforts to be responsible educators in this arena need to begin with getting this straight. Chapters 3 and 4 address this specifically and at length. I will not repeat those arguments here, except to say that dealing skillfully with evolution in the classroom cannot be done without first instructing students on what theories are, how they develop, and the central role they play in all areas of scientific research. This should begin in the Grammar School years, as described in Chapter 8.

Another important ingredient in a responsible Christian science program needs to be the cultivation of respect for all men and women of good will and their ideas. I addressed this in some detail in Chapter 1, but it bears repeating here that calling Darwin or contemporary evolutionists names is irresponsible, even slanderous. Further, Christianity is, as others have said, a "big tent."

There are many different points of view held by the 1.3 billion or so people who are Christians in this world, and one's beliefs about evolution have nothing to do with whether or not one is a genuine disciple of Jesus Christ. In fact, based on my own reading and informal queries among international friends, it seems that at least half of all the Christians worldwide accept the theory of evolution. Now, a Christian who practices infant baptism will differ with another believer who practices believers' baptism, but our approach to such differences is usually to respect the other's point of view and, when given the opportunity, to discuss the differences in order to learn and/or to persuade. We do this all the time in our Bible and theology classes. We should follow the same rule with the issue of evolution.

The parental community at my own school does not, in general, accept evolutionary theory. But, interestingly enough, one hundred percent of the people who have come to speak to me about how evolution is treated in our science classrooms have been people who want to make sure that we are not teaching young-earth creationism as scientific fact, and that the theory of evolution gets a fair treatment. Our programming for science classes needs to respect and serve the full spectrum of Christian positions on this issue.

Clearly, the majority of Evangelicals in America desire to hold the line on atheistic materialism, and many of them do so by affirming a particular interpretation of the biblical creation account and rejecting evolutionary theory. When they raise strong objections to evolutionary theory and engage in activism with local school boards, they have good intentions. But I believe that in fighting the theory like this in the social and political realms, rather than at the level of well-informed scientific inquiry and debate, they are missing a tremendous opportunity. As Christians who should be winsomely engaging the culture with the claims of the Gospel (as Paul did at the Areopagus), we should seek to build bridges between our Christian faith community and the general scientific community. Indeed, there are many Christians who are practicing scientists, for whom there is no conflict. Surveys I have heard quoted indicate that some 40% of scientists in this country are Christians. These faithful Christian servants are rightly frustrated when our discourse on this subject or our public actions and policies promote an us-them mentality between Christians and scientists.

I believe we should approach the issue in a way that fosters honest inquiry, a healthy regard for scientific research, and a spirit of constructive, intelligent dialog. Simply to stand on dogma and say to the scientific community, "You are wrong and I don't want to hear about it," shuts down inquiry and dialog and teaches our students that continued inquiry on this important issue is not necessary. How much more honoring to God would it be to teach our students that they need not fear or scorn science, that they should take every issue seriously and "study to show themselves approved," that they should help the reputation of Christians everywhere by engaging in respectful dialog, and that they should treat this issue as an ongoing arena for thoughtful inquiry (which it is) rather than a time to put their fingers in their ears?

The efforts by some in the Intelligent Design (ID) movement are a good example of constructive engagement in this issue. The notions of irreducible complexity (Behe) and specified complexity (Dembski) are good, scientifically grounded challenges to evolutionary theory that, despite the way these ideas have been mishandled by some ID people, were brought into the discussion in a very thoughtful manner. Scientists supportive of evolutionary theory have been constructively responding to these ideas at the scientific level as they should. We need more of this kind of constructive engagement by Christians.

An issue that clouds the discussion is the charge many Christian writers have made that there is an atheistic and materialistic conspiracy in the scientific community to stamp out belief in God by promoting evolutionary theory. It is obvious that many scientists are atheists. It is also obvious that some scientists (like Edward O. Wilson and Richard Dawkins) believe that science has rendered belief in God obsolete. Christians, of course, rightly reject such assertions. However, the huge majority of scientists who support evolutionary theory are not promoting such metaphysical claims, and they are not engaged in a conspiracy to stamp out belief in God. They are simply trying to conduct good scientific research and to interpret the data intelligently. It is my view that we Christians need to respect this. As I said, there are some scientists who make irresponsible metaphysical claims. But writing books that charge the entire scientific community with using the theory of evolution to subvert belief in God is also irresponsible.

Now, on to the issue itself and how we should treat it in the classroom. I will first propose some basic pedagogical principles, and then I will put forward a recommendation for a policy statement for the science department at your school.

Pedagogical Principles

In the items that follow I will be specifically addressing views held among Christians. As believers, we affirm that God is the creator of all that exists, regardless of the mechanisms he used to implement his creative purposes. I am not really concerned here with the metaphysical beliefs and claims that might be made at a secular school. Our concern is about how to be faithful in our science teaching in a Christian educational context. With that in mind, I propose that a well-structured science program will incorporate these principles:

1. *Distinguish between the naturalistic/materialistic view of evolution held by atheists and the view of evolution held by believers who accept evolutionary theory.*

    There are many people who do not believe in God at all who accept evolutionary theory. They hold that since there is no God moving the creation in any particular direction, the fundamental mechanism driving evolutionary development is undirected genetic mutation occurring strictly by accident without purpose. Views among Christians who accept evolutionary theory vary, but many argue that although the fundamental mechanism of evolution is genetic mutation, these mutations have not occurred in an undirected fashion. Rather, God works to direct the process, but does so within the mathematical boundaries of statistical probability, including what we might call "chance" or "randomness." In this way God directs creation according to his will, but his actions are (probably) mathematically undetectable. Other Christians argue that after establishing the natural laws for his creation, God allowed the creation to evolve on its own, and that its evolution could have taken many different pathways. The point here is to distinguish between a naturalistic view of evolution that leaves God out and a Christian view that places God's creative activity at the center of the issue, where it should be.

2. *Teach in a way that does not give away your own opinion on the theory of evolution.*

I am obviously proposing this principle for schools where the theory of evolution is controversial, which is the way it is at a great many Christian schools. Students should not be able to guess by your words, your demeanor or your attitude where you stand on the issue. Any teacher who will take this simple first step will introduce a level of objectivity into the classroom that will foster healthy inquiry and debate on the subject. In the same way, history teachers discuss issues in government and politics without publicizing their own political affiliations, and theology teachers discuss the Scriptures without dogmatically asserting their own denominational views. We all recognize the wisdom of such neutrality; it is the only way to function in the majority of Christian schools, which represent a plurality of opinions among the families. We need to treat the subject of evolution the same way, fostering responsible inquiry, rather than subverting it by snide remarks about how stupid this is. Responsible Christian schools wouldn't dream of tolerating snide remarks by teachers about democrats or Roman Catholicism or contemporary Christian music. We should not make snide remarks or casual negative judgments about Darwin or evolutionary theory either.

3. *Distinguish between the age of the earth issue and the issue of evolution itself.*

These two issues are related but separate, and distinguishing between them is critical because of the various distinct positions held among Christians on these issues.

The most widely held view in contemporary scientific theory holds the ages of the universe and the earth to be 13.6 billion years and 4.5 billion years, respectively. Whether these figures are approximately correct or not, the age of the earth is separate from the question of whether humans descended from other species. Evolutionary theory holds that it took several billion years for humans to evolve from the most primitive life form(s). Obviously, Christians who hold this view must also hold to an old earth perspective on the question of the age of the earth. Those in this camp embrace the view that God used, and still uses, evolutionary processes, which continue to this day, to accomplish his creative designs. It used to be common (and still is) to describe this view as "theistic evolution." However, as the debate among Christians over evolution has continued, with heightened efforts by those involved to define their terms more carefully (particularly when using terms describing scientific theories—see Chapter 4), the terms "continuous creationist" and "evolutionary creationist" have come into use to describe those who hold that God accomplishes his creative activities through evolutionary processes. A great many believing scientists hold this view.

On the other end of the spectrum are those who hold the earth to be relatively young, in the range of 10,000 to 20,000 years, and who do not accept evolutionary theory. This group holds to the principle generally referred to as "special creation of species," meaning that the species as we know them today (lizards, whales, palm trees, humans, and so on) were all specially created by God in the beginning. Third, there is a large middle group of Christians who accept the billion-year dating of the earth, but who do not accept the theory of evolution. Thus, we have three main views on this among Christians: those who hold to an old earth and theistic evolution, those who hold to a young earth and special creation of the species, and those who hold to an old earth and special creation of species.

4. *Distinguish between microevolution and macroevolution.*

When Christians engage in debates over whether evolution happened or not, what they are usually referring to is "macroevolution," a term describing the process by which species have evolved into other species millions of times over biological history. This is the theoretical proposition that is so controversial among Christians. Microevolution, on the other hand, happens all the time and is not questioned by anyone. Microevolution is basically variation within a species. When humans breed dogs, or cows, or roses, or peaches they are leveraging microevolutionary processes. No matter how far the breeding process goes, a dog is still a dog, and a dog breeder is not going to convert his dogs into anything else.

It is worth noting that some of Darwin's early scientific observations were about moths changing color to adapt to their environment. These are microevolutionary changes. It is also worth noting that the scientific community tends to shy away from these terms these days because they have been used for so long by Christians to force a wedge into evolutionary theory. However, given the fact that the debate over evolution continues among Christians, the terms are still useful.

5. *Distinguish between the creation of life and the evolution of life (speciation).*

   Rightly explained, evolutionary theory does not propose a mechanism for the creation of life, only for its development. There are scientific theories that attempt to explain the origin of life, but these theories are far more speculative and less grounded in scientific evidence than the theory of evolution. It is certainly possible to accept the validity of evolution without accepting a naturalistic explanation for the creation of life.

6. *Recognize the historical significance given to the Big Bang theory.*

   When Edwin Hubble first proposed his theory that the universe is accelerating outward from an ancient initial explosion, now known as the Big Bang, his ideas were broadly recognized as implying the existence of God. The reasoning was that if the universe was static, then it could have been the way it is eternally without the need for a creator. But if it had a *beginning*, then some agent had to be its initial cause. Thus, the Big Bang implied the existence of the One, presumably God, who made it happen. To many believers it still implies this, even though the Big Bang cannot be regarded as proof for the existence of God. It is one of the ironies of scientific history that the Big Bang, originally regarded as strong evidence for God's existence, is now regarded by many Christians as an unbiblical idea.

7. *Treat evolutionary theory as all other theories are treated—as a proper scientific theory.*

   A common Christian strategy to undermine the credibility of evolutionary theory is to outline the "strengths and weaknesses" in the theory, or to note the "gaps" in the theory. No one does this with any other theory, quantum theory, for example, with its bizarre implications about reality and (as Einstein put it) "God playing dice with the universe." But it is commonly done with evolution. I submit that this is just a bad strategy for dealing with this subject. Without doubt, the question of whether or not we evolved from primitive species will some day be resolved to nearly everyone's satisfaction and Christians won't be arguing about it any more. At some point the scientific evidence either for evolution or against it will be so compelling that no reasonable person will dispute it. The general scientific community, Christian and non-Christian alike, has already passed this point, and regards evolution as a theory that is just as firmly established as other major theories.

Right now we are in the same position as those during the Copernican Revolution, when Church theology aligned itself with one particular view, and when a new different view came along everyone simply had to study the matter as carefully as possible for a hundred years or so until everything became clear. It did not help matters that the Church at that time persecuted those who accepted the new views about the solar system. Similarly, it does not help now to treat evolution as an enemy to be fought with every last breath. As I've written, many Christians, including thousands of Christian scientists, accept evolutionary theory, and these people are not the enemies of those who don't. The real philosophy that needs to be opposed is naturalistic materialism—the view that there is no God and that everything got here by meaningless accident. In my view, instead of using our biology courses as an opportunity to study "strengths and weaknesses," there are other more interesting questions that can be asked which will show our students how to engage issues like this thoughtfully and intelligently, such as:

a. *What is the well-established evidence for evolution?*
b. *What does it mean when scientists state that the entire field of molecular biology depends on evolution?*
c. *What are the scientific challenges that have been raised against evolution, such as Intelligent Design, and what can be said about them (see below)?*

d. *What are the reasons, based on biblical texts or otherwise, why many Christians oppose evolutionary theory?*

e. *If we suppose that evolution happened (and, presumably, is still happening), what does this imply about the biblical texts, and are there other ways to interpret these texts that are consistent both with evolutionary theory and orthodox Christian theology?*

In summary, our classrooms should be structured to promote responsible inquiry. This is what we do, or should be doing, in every other discipline. We teach our students how to read the revelation given to us in the Scriptures, and then we challenge them to apply it to the difficult questions we face in every discipline—war, capital punishment, money, sexuality, biomedicine, law, and on and on. All these disciplines entail hosts of difficult and challenging issues that involve disagreements among intelligent Christians of good will. Schools that try to characterize these issues as having obvious and straightforward answers, and demonizing anyone not embracing these obvious answers, are guilty of propagandizing. Responsible schools will avoid this and instead teach their students to use the minds God gave them to reason, to discern, and to not be afraid of seeking and proclaiming the truth.

8. *Teach and explain the details of evolutionary theory.*

It would be irresponsible to send our students off to college from our schools without telling them about the views held by the vast majority of the scientific community regarding the origin of the human species. Regardless of whether we accept the theory of evolution or not, most scientists do and our students are going to be in their classes in college. To avoid teaching the subject because one disagrees with it is equivalent to avoiding Plato because one dislikes his views on government, or Ernest Hemingway because one dislikes his immorality. The fact is that Plato's philosophy, Hemingway's fiction, and evolutionary theory are all enormous contributors to western culture as it is today. Responsible educators must assure that their students are conversant with all such major contributors to our intellectual tradition.

9. *Discuss the challenges to evolutionary theory posed by Intelligent Design.*

Our goal is for our students to understand the issue of evolution, why there is a controversy, and how we can think intelligently about it in such a way as to honor God. In order to do this they need to understand what Intelligent Design is and what it is not.

Intelligent Design is not an alternative theory to evolution. Intelligent Design is a challenge to evolutionary theory and, separately, to philosophical naturalism (which holds that there is no God and things got here by meaningless accident). More specifically, the Intelligent Design movement has mounted a specific challenge to the adequacy of genetic mutation (the fundamental evolutionary mechanism) to accomplish what most scientists theorize it has accomplished (the evolution of the species), and a specific challenge to naturalistic materialism. I will briefly summarize these two ideas.

The challenge to the adequacy of genetic mutations as the evolutionary mechanism is what Michael Behe calls "irreducible complexity" in his book *Darwin's Black Box*. Behe notes that there are many microscopic, molecular systems in our bodies that operate like little machines with many parts, and, like a basic mousetrap, these machines must have all of their parts present simultaneously in order to function at all. Everyone agrees that there are many such "irreducibly complex" biological systems at the molecular level. That is, that there are many such systems that must have all of their parts present to work properly. However, Behe also uses the phrase "irreducible complexity" to argue that since, according to evolutionary theory, each part would presumably have evolved one genetic mutation at a time, there would be no way for them to do so, since the advantage the organism gains by possessing these mutations is not realized until the entire machine is present. This is an important and provocative argument and our students should be able to articulate it and give examples of such biological machines.

Behe's book was published in 1996. Since then several important publications have directly refuted Behe's claim that there is no evolutionary mechanism by which these molecular machines could have evolved. Our students should be familiar with these arguments as well. Kenneth Miller's 1999 book *Finding Darwin's God* was the first. Miller's book is very readable and includes an extended Chapter ("God the Mechanic") directly challenging Behe's thesis. Keith Miller's 2003 publication *Perspectives on an Evolving Creation* is an anthology of articles by quite a few Christian scientists and theologians. The articles in this book tend to be quite technical, and thus are challenging reading, but the book contains important contributions on this issue. The basic idea described in each of these publications is that these irreducibly complex molecular machines did not need to evolve from scratch one piece at a time. Instead, biological systems use explainable natural processes to seize and adapt various existing components from other systems (components that were present to serve completely different purposes), and utilize them to perform novel functions.

The challenge to naturalistic materialism, known as "specified complexity," was most famously made by William Dembski in his 1999 book *Intelligent Design*. Dembski argues in this way. If one dropped a Scrabble game on the floor and the three letter sequence C-A-T came up, well, that could happen because that three letter sequence is not very "complex," and it is not improbable that it would happen. The letters do have meaning for us, having to do with cats, which Dembski calls "specification." On the other hand, if the letters A-F-G-H-E-D-H-U-F-S-S-S-F came up, this sequence has no meaning, no specification, but it is very complex and the probability of it happening is extremely small. Dembski puts these together and argues that the biological systems we see today have both specification and complexity, and the only way this can happen is if they were designed by an intelligent being. This is a powerful argument against the notion that evolution happened by itself—intelligence can only come from intelligence, or as Dembski puts it, meaningful information can only come from an intelligent source. But Christians who accept evolutionary theory are not promoting accidental, meaningless evolution, but purposeful, directed evolution. Some Christians hold that God continuously infuses his intelligence and creativity into the creation all along the way as the creation goes on evolving. Others hold that he gave the universe a set of instructions to follow and turned it loose, knowing that several different evolutional outcomes were possible. This is a very interesting issue to discuss, and believers of good will hold various points of view. But the issue here is that the presence of specified complexity in nature is an argument for an ultimate divine agent having been the cause of creation, not an argument about whether evolution could have happened at all.

10. *Do your reading.*

I am constantly amazed at the number of Christian science teachers I meet who have strong opinions on this issue, and who are charged with teaching on it one way or another, but who have not read the basic texts involved. Surely that is like a teacher in a Government class who has not read the U.S. Constitution or the *Federalist Papers.* If you are involved with this issue as a science teacher, you need to read the basic books. Just read one or two of them per year and in a few years you will be vastly more informed and better equipped to lead your classes. Start with Darwin and read Behe, Dembski, Collins, and others. If one includes all the books by Philip K. Johnson and everyone else, there is a lot to read. Moreover, new books are published all the time by Christians and non-Christians alike and the literature is now massive. I recommend that all Christian science teachers begin with this basic canon and go from there:

a. Charles Darwin, *On the Origin of Species*
b. Michael Behe, *Darwin's Black Box*
c. William Dembski, *Intelligent Design*
d. Miller, Kenneth, *Finding Darwin's God*
e. Miller, Keith, ed., *Perspectives on an Evolving Creation*
f. Collins, Francis, *The Language of God*
g. Haarsma, Deborah and Haarsma, Loren, *Origins: A Reformed Look at Creation, Design and Evolution*

A Recommended Policy Statement for Your Science Department

Below is a single-page policy statement that may be used as a guideline for the science department at a Christian school:

The Role of Evolution in the Science Department
Present Realities

1. Christians are deeply divided on the subject of evolution. Worldwide, it appears that most Christians accept evolution; in America, Evangelicals often strongly oppose the theory of evolution, either because they believe the theory conflicts with Scripture or because they associate it with naturalistic materialism.
2. The subject of evolution is often confused with related issues such as the age of the earth and the big bang. Further, the creation of life is a separate issue from the evolution of life (speciation).
3. People often use inaccurate language when discussing the issue, creating further confusion. Terms such as fact, theory, and science are frequently misused.
4. The theory of the evolution of all life by natural selection through the mechanism of genetic mutation, while not universally accepted in the scientific community, is the prevailing view. Evolution is widely understood to be the foundation for the discipline molecular biology, and of critical importance for the disciplines of physiology and medicine.
5. Scholars, graduate students and professors who oppose evolution prior to obtaining tenure are very susceptible to censure in academia.

Goals of the Biology Program vis à vis Evolution

1. Students can articulate the basic tenets of evolutionary theory using correct terminology, including:
   a. Natural Selection (survival of the fittest)
   b. Variation
   c. Genetic modification by mutation
   d. Speciation
   e. The fossil record
2. Students can describe the current diversity of views among Christians and non-Christians.
3. Students can articulate the distinction between the age of the earth/universe issue and the issue of evolution itself.

4. Students can articulate examples of scientific evidence supporting evolutionary theory.
5. Students can use Scripture to explain why many Christians reject the theory of evolution, and can articulate the challenges to evolution proposed by the Intelligent Design movement.

Principles of Our School's Presentation

1. We endorse neither evolution nor special creation of species as scientifically factual. Instead, we regard this entire subject as a fascinating field of scientific inquiry. We strive to train students to be teachable, thoughtful, responsible participants in the debate. We do not engage in a "teach the controversy" approach, or a "strength and weaknesses" approach, which some Christian schools deliberately use to undermine evolutionary theory.
2. We hold that there is no conflict between Christian faith (the Bible) and science when both are rightly understood. Moreover, we affirm and advance the convictions outlined in our school's Statement of Faith. In particular, we affirm that God is the Creator of all things, regardless of the processes or methods he used to bring the creation to its present state of complexity.
3. We maintain that our faith informs our science, and vice versa.
4. We expect that eventually the controversy will be resolved as inquiry continues, but we acknowledge that both the Church and the scientific community have been wrong before on such questions, and that the resolution of this question could, in principle, go either way.